Studies in Computational Intelligence

Volume 507

Series Editor

J. Kacprzyk, Warsaw, Poland

For further volumes:
http://www.springer.com/series/7092

Anis Koubâa · Abdelmajid Khelil
Editors

Cooperative Robots and Sensor Networks

Springer

Editors
Anis Koubâa
Prince Sultan University
Riyadh
Saudi Arabia

Abdelmajid Khelil
DEEDS Group, CS Department
TU Darmstadt
Darmstadt
Germany

ISSN 1860-949X ISSN 1860-9503 (electronic)
ISBN 978-3-662-51260-9 ISBN 978-3-642-39301-3 (eBook)
DOI 10.1007/978-3-642-39301-3
Springer Heidelberg New York Dordrecht London

Printed on acid-free paper

Springer is part of Springer Science+Business Media (www.springer.com)

Preface

Mobile robots and Wireless Sensor Networks (WSNs) have enabled great potentials and a large space for ubiquitous and pervasive applications. Robotics and WSNs have mostly been considered as separate research fields and little work has investigated the marriage between these two technologies. However, these two technologies share several features, enable common cyber-physical applications, and provide complementary support to each other. The primary objective of the book is to provide a reference for cutting-edge studies and research trends pertaining to robotics and sensor networks, and in particular for the coupling between them.

The book consists of five chapters. Chapter 1 presents a cooperation strategy for teams of multiple autonomous vehicles to solve the rendezvous problem. Chapter 2 is motivated by the need to improve existing solutions that deal with connectivity prediction, and proposed a genetic machine learning approach for link-quality prediction. Chapter 3 presents an architecture for indoor navigation using an Android smartphone for guiding a variety of users, from sighted to the visually impaired, to their intended destination. In Chapter 4, the authors deal with accurate prediction modeling of ocean currents for underwater glider navigation. In Chapter 5, the authors discuss the challenges and limitations of RSS-based localization mechanisms and propose, EasyLoc, an autonomous and practical RSS-based localization technique that satisfies ease of development and implementation.

Contents

Reviewers

Anis Koubâa, Al-Imam Mohamed bin Saud University (Riyadh)/CISTER Research Unit (Porto), aska@isep.ipp.pt, http://www.dei.isep.ipp.pt/~akoubaa

Abdelmajid Khelil, TU Darmstadt, abdelmajid.khelil@huawei.com, http://www.deeds.informatik.tu-darmstadt.de/khelil/index.html

Raul Aquino, University of Colima, aquinor@ucol.mx, http://raquino.siteldi solutions.com/

Luis Almeida, Fac. de Eng. da Universidade do Porto, lda@fe.up.pt

Adel Ben Mnaouer, University of Trinidad and Tobago, adel_mnaouer@yahoo.com, http://www.cud.ac.ae

Andrea Zanella, Dep. of Information Engineering (DEI), University of Padova, zanella@dei.unipd.it, http://dgt.dei.unipd.it/people/read/Andrea+Zanella/

Habib Youssef, Institut Superieur d'Informatique et des Technologies de Communication de Hammam Sousse, habib.youssef@fsm.rnu.tn, http://www.infcom.rnu.tn/

Jiong Jin, The University of Melbourne, jiong.jin@gmail.com

Gian Pietro Picco, University of Trento, gianpietro.picco@unitn.it

Daniel Mosse, Univ of Pittsburgh, mosse@cs.pitt.edu

Andreas Willig, University of Canterbury, awillig@ieee.org

Joerg Haehner, Leibniz Universitaet Hannover, joerg.haehner@informatik.uni-augsburg.de, http://www.informatik.uni-augsburg.de/lehrstuehle/oc/

Marco Zuniga, University of Duisburg-Essen, mzunigaz@gmail.com

Fumin Zhang, fumin@gatech.edu, http://users.ece.gateh.edu/~fumin

Silvia Santini, Darmstadt Univ, santinis@wsn.tu-darmstadt.de, http://people.inf.ethz.ch/santinis/

Rongxing Lu, University of Waterloo, rxlu@bbcr.uwaterloo.ca, http://www.ntu.edu.sg/home/rxlu/

Ye-Qiong Song, LORIA-Nancy University-INPL, Ye-Qiong.Song@loria.fr, http://www.loria.fr/~song/

M. Bernardine Dias, Carnegie Mellon University, mbdias@ri.cmu.edu

Mário Alves, CISTER Research Unit, Politécnico do Porto, mjf@isep.ipp.pt, http://www.cister.isep.ipp.pt

Carlos Sagues, Universidad de Zaragoza, csagues@unizar.es, http://webdiis.unizar.es/~csagues/

Geoffrey Hollinger, Carnegie Mellon University, geoff.hollinger@gmail.com

Xianghui Cao, edwincxh@gmail.com

Tian Huang, th@eng.warwick.ac.uk, http://http//www.tju.edu.cn

Xu Li, University of Waterloo, easylix@gmail.com, https://ece.uwaterloo.ca/~x279li/

Michel Devy, LAAS-CNRS, Michel.Devy@laas.fr, http://homepages.laas.fr/michel

Mohamed Elarbi-Boudihir, IMAM, elarbi-boudihir@lycos.com

Yuuichi Teranishi, Osaka University, teranisi@ist.osaka-u.ac.jp

Naoki Wakamiya, Osaka University, wakamiya@ist.osaka-u.ac.jp, http://www.anarg.jp/~wakamiya/

Fakir Dawood, fakir.dawood@yuc.edu.sa

Takashi Tsubouchi, tsubo@esys.tsukuba.ac.jp

Ramiro Martinez, University of Seville, jdedios@cartuja.us.es

Maissa Ben Jemaa, University of Sfax, Tunisia, maissa.benjamaa@rtrackp.com

Chapter 1
Genetic Machine Learning Approach for Link Quality Prediction in Mobile Wireless Sensor Networks

Gustavo Medeiros de Araújo, A. R. Pinto, Jörg Kaiser and Leandro Buss Becker

Abstract Establishing adequate RF (Radio Frequency) connectivity is the basic requirement for the proper operation of any wireless network. In a mobile wireless network it is a challenge for applications and protocols to deal with connectivity problems, as links might get up and down frequently. In these scenarios, having knowledge of the node remaining connectivity time can avoid unnecessary or even unuseful control/data messages transmissions. The current paper presents the so-called *Genetic Machine Learning Approach for Link Quality Prediction*, or simply GMLA, which is a solution to forecast the remainder RF connectivity time in mobile environments. Differently from all related works, GMLA allows building

G. Medeiros de Araújo (✉) · L. B. Becker
Department of Automation and Control Systems, Federal University of Santa Catarina, Florianópolis, Brazil
e-mail: araujo@das.ufsc.br

L. B. Becker
e-mail: lbecker@das.ufsc.br

G. Medeiros de Araújo · L. B. Becker
UFSC/CTC/DAS/PPGEAS, Room 214, Trindade, PO 476, CEP 88040-900, Florianópolis, Brazil

A. R. Pinto
Department of Computer Science and Statistics, Paulista State University (UNESP), São Paulo, Brazil
e-mail: arpinto@ibilce.unesp.br

A. R. Pinto
Rua Cristóvão Colombo, 2265 - Jardim Nazareth, CEP 15054-000, São José do Rio Preto, SP, Brazil

Jörg Kaiser
Department of Distributed Systems, Otto-Von-Guericke-Univesität Magdeburg, Magdeburg, Germany
e-mail: kaiser@ivs.cs.uni-magdeburg.de

Jörg Kaiser
Universitätsplatz, 2 D-39106, Room 323, Magdeburg, Germany

A. Koubâa and A. Khelil (eds.), *Cooperative Robots and Sensor Networks*,
Studies in Computational Intelligence 507, DOI: 10.1007/978-3-642-39301-3_1,

connectivity knowledge to estimate the RF link duration without the need of a pre-runtime phase. This allows to apply GMLA at unknown environments and mobility patterns. Its structure combines a Classifier System with a Markov chain model of the RF link quality. As the Markov model parameters are discovered on-the-fly, there is no need of a previous history to feed the Markov model. Obtained simulation results show that GMLA is a very suitable solution, as it outperforms approaches that use geographical positioning systems (GPS) and also approaches that use link-quality prediction, such as BD and MTCP. GMLA is generic enough to be applied to any layer of the communication protocol stack, especially in the link and network layers.

Keywords Mobile wireless networks · RF connectivity prediction · Classifier systems

1 Introduction

Recently, there has emerged an eminent need for using Wireless Sensor Networks (WSN) in scenarios dealing with mobility. A typical scenario could be robots moving through a factory while communicating with other robots and also with fixed nodes to perform collaborative work [6]. Another example is the collaboration between WSNs and Unmanned Aerial Vehicles (UAVs), which aim at location awareness and target tracking [10, 13, 27].

However, deploying such mobile wireless networks is not an easy task [11]. One of the problems that is of special interest in this paper concerns the radio frequency (RF) connectivity, since establishing proper RF connectivity is the basic requirement for the proper operation of any wireless network. In mobile applications, network protocols are challenged to deal with RF connectivity problems, as links might get up and down frequently. In these scenarios, having knowledge of the node remaining connectivity time can avoid unnecessary or even unuseful control/data messages transmissions. This property is particularly useful for network and link layer protocols.

The awareness of connectivity variations in mobile environments is typically handled by employing either location-based or data training strategies. While location-based strategies require some kind of location system, data training strategies rely on data gathered from the network card (see [1, 9]). Both strategies ensure building some sort of connectivity knowledge. In the next Section a more detailed discussion about these strategies will be presented, including the related theoretical models.

Differently from all related works, the solution presented in this paper allows building a connectivity knowledge in order to estimate the link duration without the need of a pre-runtime (design) phase. It can perform the connectivity prediction while the system is already executing (this mechanism is known as "on-the-fly"). The proposed solution is called *Genetic Machine Learning Approach for Link Quality Prediction*, or simply GMLA. To be able to perform on-the-fly, GMLA makes use of an evolutionary approach that learns the mobility pattern along the first seconds of the algorithm execution. GMLA is generic enough to be applied to any layer of the communication protocol stack, especially in the link and network layers.

The remaining sections of this paper is organized as follows: Sect. 2 provides a discussion about some fundamental aspects related with connectivity estimation; Sect. 3 describes some works related with the connectivity prediction problem. Section 4 details the Oriented Birth-Death (OBD) model, which serve as basis for our proposal; Sect. 5 presents our proposed GMLA; Sect. 6 describes GMLA evaluation and discusses the obtained results; finally; Finally, in Sect. 7 we present the final remarks and discuss possible applications for GMLA.

2 Connectivity Concerns

This Section highlights some issues that concern all approaches that deal with connectivity prediction. Such issues can be organized in four different aspects: (i) The need of a positioning system, (ii) The need of a history data log, (iii) The difference between link quality estimator and link quality predictor, and (iv) The type of model used to perform the connectivity estimation.

The first aspect includes approaches that need some kind of positioning system, such as GPS or GPS-free. The need of a GPS has some drawbacks, like increasing the costs in large scale implementations and higher energy consumption. Moreover, it is only suitable for outdoor scenarios. GPS-free systems overcame some of these limitation, as they were designed to be used with indoor scenarios. However, they require extra antennas such as *Time Difference of Arrival* (TDOA), or devices like Cricket Compass [22] or even anchor nodes. Anchor nodes represent fixed nodes that have their position known beforehand. They broadcast their position by flooding the network with this information, so that other mobile nodes can calculate their own position using triangulation.

The second aspect concerns all approaches that need a history data log. Such data log could be of many different variables, such as coordinate position or link quality (e.g. RSSI or SNR). The history data log is normally recorded during an off-line mode to be processed and used later. The approaches that need data history are pattern representations contained in the history to fill the model. Thereby, after off-line processing time, it is expected to have the same behavior in the collecting time. The main drawback of this approach relates with possible pattern modifications. If this occurs, a new data history is required.

The third aspect is the difference between link quality estimator and link quality predictor, also including its relation in the context of wireless sensor networks. A link quality estimator (LQE) is used to estimate the reliability of the wireless channel. LQE calculation can be performed by the wireless card resulting in information like RSSI, SNR, and LQI (Link Quality Indicator). It can also be represented by the packet reception rate (PRR), which is performed as the packet is received. On the other hand, a link quality predictor (LQP) estimates the future value from the link quality. In other words, it tries to guess the future state of the wireless link.

The fourth and last aspect relates with the model chosen to estimate the connectivity. There are three main models that are largely used to perform connectivity predictions: (i) Markov Models, (ii) Time Series, and (iii) Relative Speed. The Markov Model and Time series need the data history log to compute the estimation. Finally, the Relative Speed models need an accurate positioning system. The main drawback from Markov Model is that, depending on the process, the number of states could be very large. For instance, if the state is considered as a node and the scenario has a large number of nodes, the model size may became so big that it may bring problems to resources constrained devices. However this depends on how the problem is modeled. Time Series may have the same disadvantage, as the entire data history must be loaded in the memory to be processed. The use of Relative Speed requires a positioning system to give the coordinate information in order to calculate the vector velocity from each node. The limitations of Relative Speed models relates to the need of positioning systems, as previously discussed.

Next we present the main works found in the literature about connectivity prediction.

3 Related Works

The work presented in [25] was one of precursors in predicting the connectivity between nodes in mobile environments. The authors presented a deterministic model to predict connectivity (see Eq. 1). The proposed equation can give the remaining connection time between mobile nodes. The adopted information set is composed by the knowledge of position (provided by a GPS), angle θ, velocity v, and the signal range r from nodes.

$$Dt = \frac{-(ab+cd)+\sqrt{(a^2+c^2)r^2-(ad-bc)^2}}{a^2+c^2} \tag{1}$$

where Dt is the amount of time two mobile nodes will stay connected and

$$a = v_i cos\theta_i - v_j cos\theta_j$$
$$b = x_i - x_j$$
$$c = v_i sin\theta_i - v_j sin\theta_j$$
$$d = y_i - y_j$$

In [23], the authors developed a technique to predict link disconnections targeting collaborative applications in MANETs. They assumed that all nodes have their position provided by GPS-free system. At startup time, all nodes must be connected to build a graph (by the way, this can be considered a limitation of this approach). The weight of each edge represents the distance between neighbor nodes. The variation on this distance is used to build the prediction graph. With such graph the coordinator node can foresee when a node will disconnect and might try to avoid it.

In [7] the proposal of the authors is to measure the duration of links (or link stability) among nodes. The adopted prediction scheme combines the knowledge of position with channel fading information. First, they use knowledge about node mobility to predict the link lifetime. Then they combine the link lifetime with the fading channel statistics to obtain the link duration. To calculate the link durability, they assume that all nodes use a GPS to provide their locations and speeds. Therefore, they combine the knowledge of relative speed with statistics from channel fading to forecast link duration.

A model that allows making better use of the communication resources from a WLAN by taking into account the user mobility history is presented in [20]. Such history is used as input to a Markov Model, that aims to predict the node mobility. The approach requires positioning information, so a GPS device is needed. This scheme is based on localization prediction, as the nodes and routers have their positioning knowledge, the model is built based on the probability of changing the localization. For instance, if a node is moving from one area to another the routers can foresee the change in the node localization and anticipate bandwidth reservation. The main drawback of such approach is the need of a localization system.

In a different approach, the authors of [12] use the knowledge of link quality to predict the connectivity between nodes. They used a Time Series to model the changes on the link quality along time while the modes are moving. This approach requires history. In a predetermined scenario the history of user mobility was used as input parameter in order to build the model. For this reason, it was not necessary to use any additional hardware to get positioning information.

The MTCP approach presented in [24] represents RSSI variations using a Markov model with five states. Their intention is to use MTCP to improve the backoff mechanism from the MAC protocol, so that the nodes can save energy and also enhance the available bandwidth. Obtained results are very positive considering one step ahead prediction.

The Birth-Death (BD) Markov Model presented in [15] has the advantage of avoiding the use of positioning information. Instead, it predicts the quality of wireless links in WiFi networks using link quality information by means of the *Signal-to-Noise Ratio* (SNR). The main limitation of this model is the fact that it does not properly represent the movement of the nodes. It ignores the principle of inertia, which states that an object cannot just change its direction immediately. Such problem is expressed by the proposed Markov Model, which suggests that a node has the same probability to either increase or decrease the signal strength, i.e., to maintain or change its direction. However, in fact, the probability that the node keeps going in the same direction is much higher than of changing its direction. This problem was overcome by our proposed Oriented Birth-Death (OBD) [3], which is detailed in the next Section.

Analyzing all related works, one can see that they use either Markov Models or Time Series to represent the prediction mechanism. Moreover, it is observed that all solutions which require a localization device (like a GPS) have restrictions to be applied in indoor scenarios. Also, anchor nodes are needed with the

GPS-free proposals. In both cases, geographic location awareness is used, increasing the computational complexity of the solution and making them less flexible.

Also, some solutions also have strict model restrictions. For example, the model in [23] is dependent on the number of nodes, since each node is represented by a state in the Markov chain. This may lead to scalability problems, as it might be hard to handle a high number of nodes. Furthermore, it is not suitable for applications with variable number of nodes. The drawback from [12] is that it does not address future links, that is, if a new neighbor node appears the model cannot forecast the connectivity with it. Therefore, if mobility pattern changes the model will be out of date and a new off-line training must be performed.

4 Oriented Birth-Death Model

The ODB [3] model was developed to overcome the lack of orientation of the BD model [15]. To provide a better understanding about this issue we present some measurements containing the SNR variation while a ground robot is moving around a room, sending messages to a static sensor node located in the corner of the room. Such results are presented in Fig. 1. The line with perturbations (lighter color) represents the raw SNR and the more stable one (darker color) represents the SNR filtered with a Kalman Filter. In fact, the more stable curve represents the robot mobility pattern. When the robot approaches the sensor node, the RF signal strength reaches its peak. Similarly, when the robot departs from the sensor node, the RF signal strength decreases until there is no power. Approaching and departing are patterns that exist in all types of mobility.

Just like the BD model, OBD also uses a Markov chain model to predict the link quality in the future. Markov models have the advantage that they do not need to keep the history in memory, which is a good point to resource constrained devices like a sensor node. In fact, the history is already embedded in the states transitions probability of the Markov model. Such probabilities are stored in a fixed-size matrix.

The main difference between O-BD and BD relies on the fact that the former takes the nodes movement orientation/direction into consideration. This does not mean that a node cannot change its direction. What is means is that this has a very low probability to happen. To represent such issue, all intermediate states of the automaton are duplicated, as shown in Fig. 2. The states on top represent nodes in one direction, and the state on the bottom represent nodes in the opposite direction. The states on the extreme are not duplicated because it only makes sense for them to go towards a single direction. The two features of approaching and departing, which can be seen in Fig. 1, are presented in O-BD.

The knowledge provided by the model is helpful to allow forecasting the next state. Therefore we used the discrete Markov model theory, where each transition happens within a second. The description about how Discrete Markov Chain works to estimate the future state is described in next Subsection.

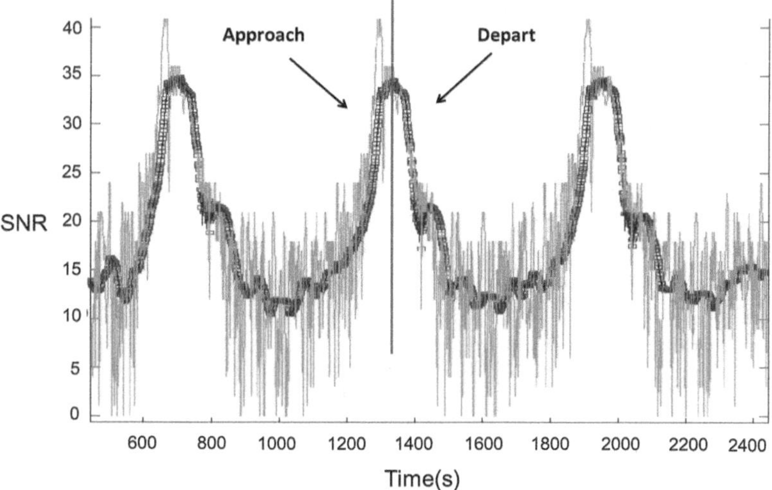

Fig. 1 SNR representing approaching and departing situation over the time

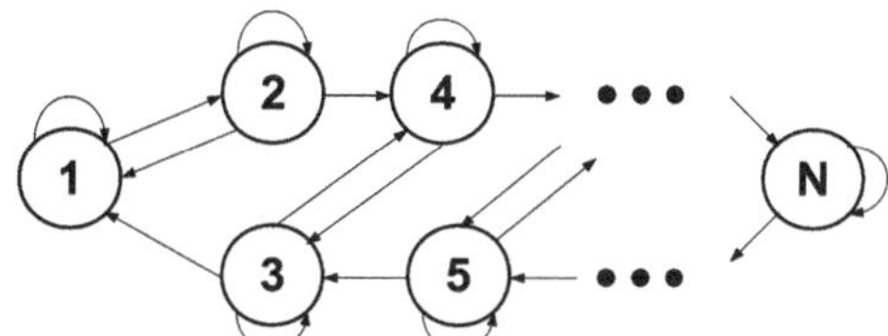

Fig. 2 Oriented birth-death automaton

Discrete Markov Chain The discrete Markov chain could be described as a triple $< S, T, \pi >$. S represents the set of states, as illustrated in Fig. 2, and T is the transition matrix that represents the probability to go from a certain state i to the state j. It is related to each possible scenario (mobility pattern), and it can be obtained by means of simulation or practical experimentation. Finally, π is the initial probability distribution of the set S. To choose the next state we use the vector state probability, as shown in Eq. 2. T^n is the transition matrix powered by n (number of steps ahead). After this calculation, it is possible to obtain the new probability vector π^n and finally choose the next state using Eq. 3.

$$\pi^n = \pi * T^n \tag{2}$$

$$s^n = Max[s_1^n, s_2^n, \ldots, s_k^n,] \tag{3}$$

Similarly to all models which are based in a history, the discrete Markov chain is affected by modification of the mobility pattern. In the next Section we present an approach that is capable to overcome these limitations.

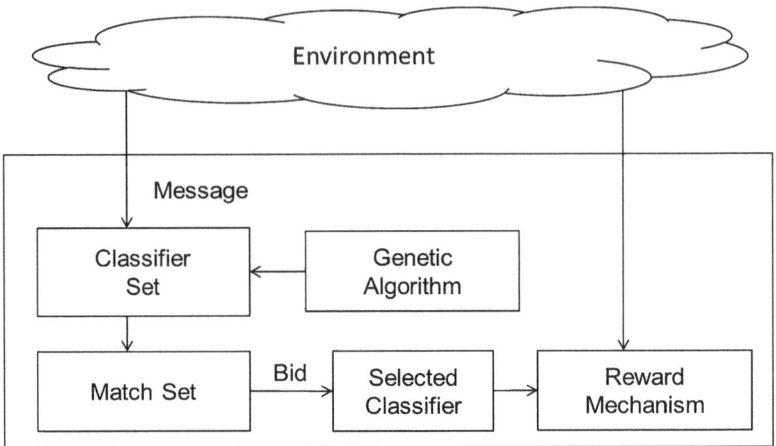

Fig. 3 Classifier system scheme

5 GMLA Presentation

The Genetic Machine Learning Approach for Link Quality Prediction (GMLA) is based on Classifier Systems [14], a machine-learning technique based on genetic algorithms (GA) that is capable of learning syntactically simple rules. It is used to discover at runtime the parameters of Markov model that represent the RF quality. Fig. 3 presents an overview of the proposed approach, which is detailed in the incoming Subsections.

5.1 Classifier Systems

A message received from the environment can activate one or more classifiers. As classifiers are selected, they perform their rules. Afterwards, the selected classifiers are rewarded based on their performance. GAs consider a population of classifiers for some optimization problem. In this case there are individuals (classifiers) representing their genotypes, which are usually a set of bits or characters. This population is evolved by the GA after a predetermined number of consults. At each generation of answers, a new set of artificial creatures (classifiers) is generated. The answers are based on fragments of the most adapted previous individuals.

The main focus of a GA is its robustness. If the system is more robust, it requires a smaller number of interventions or redefinitions. Moreover, it will achieve higher levels of adaptation and will be able to perform better and last longer. The main difference between a classical GA approach and a classifier system is that the latter just evolves its population after some consults to the classifier set. Thus, it will perform the evolution process while it is running.

5.2 Genetic Machine Learning Approach

The proposed *Genetic Machine Learning Approach* (GMLA) is an extention of OBD model [3], as it also uses a Markov Chain model to predict the link quality in the future.

As mentioned, the transition matrix T used by the Markov models is dependent on the environment characteristics, like for example the mobility pattern. To overcome this issue and allow the same protocol-stack to be used in unknown mobility conditions, we developed an evolutionary algorithm that is capable to find the best T on the fly. For π, we assumed that each state has the probability *1/m*, where m is the number of states of the model.

A node may begin to communicate at any signal range. The time needed for training the model and finding the best T varies with the amount of link quality samples available. Of course this depends on each application scenario. As the link quality samples arrive with higher frequency, the faster the model can be built. The opposite is also true. However, it is not always the case that this learning phase is required. For instance, nodes could receive T from other nodes that already learned, or even from the environment itself.

Each chromosome in GMLA represents a transition matrix and each gene represents a row, which is the state in the automaton. Figure 4 represents the idea of a chromosome. As each gene is a row, it contains the values from transitions probabilities to go from the state i to j. As each gene is a row containing the transitions probabilities, it must guarantee the basic role for Markov model, which is the sum of transitions probabilities from a state must sum 100 %. To ensure this property, each gene executes the Algorithm 1. For instance, if the gene has three transitions, the first transition randomly selects a number N_1 between 0 and 100, then the second transition randomly selectes a number N_2 between 0 and $(100 - N_1)$ and the last one takes the rest, $N_3 = 100 - (N_1 + N_2)$. In the end, the sum of $N_1 + N_2 + N_3$ is 100 %. In this way, each chromosome is created and represents the model from automaton 2.

During startup each individual is randomly generated, then it evolves according to the phases shown in Fig. 5. After sampling the link quality, the individuals map the link quality value with the current state. The next stage, they guess the next state for N transitions ahead by computing Eqs. 2 and 3. After N steps further, all individuals are evaluated. The link quality is sampled and mapped again, then it is compared with the guessed state. All individuals that guessed correctly are rewarded. This process continues until the number of queries finishes.

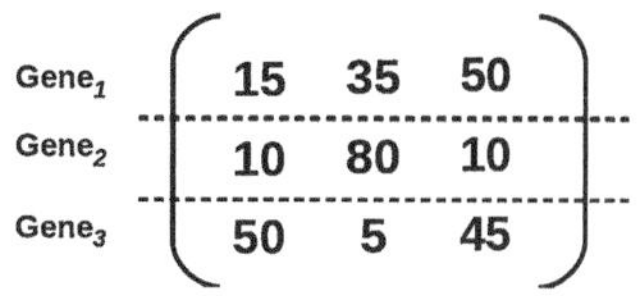

Fig. 4 The chromosome as a matrix and the gene as a row

$Gene_1$	15	35	50
$Gene_2$	10	80	10
$Gene_3$	50	5	45

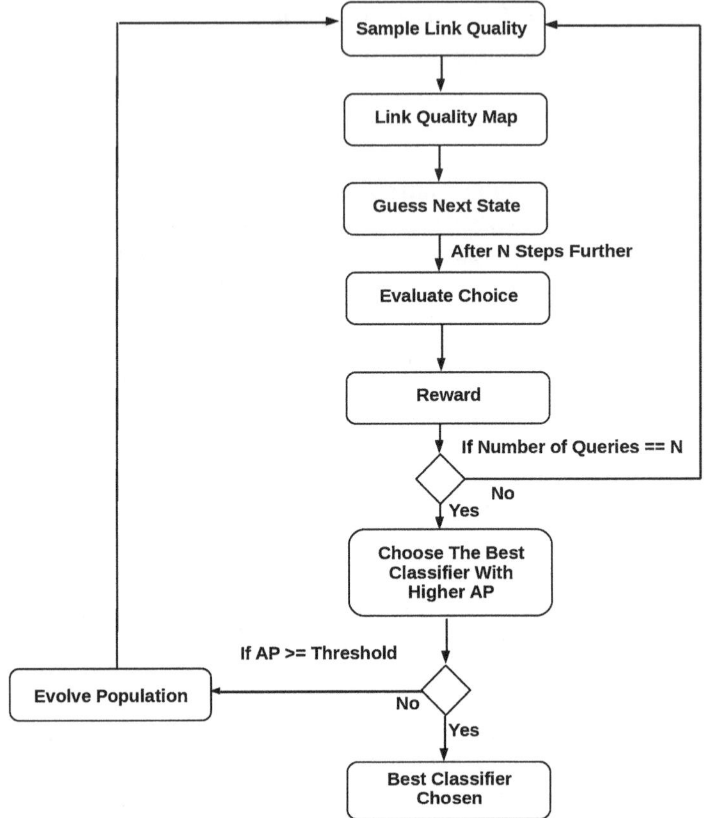

Fig. 5 GMLA—genetic machine learning approach

After a certain number of queries, the next stage is to choose the best classifier, by applying the fitiness function (Eq. 4). The accuracy percentage (*AP*) is the number of correct guesses P_t divided by the number of queries Q. The classifer with higher *AP* is chosen. If the best classifier's *AP* is higher or equal to a determined threshold, the GMLA process finishes. However, if the best classifier's *AP* is lower than the determined threshold, all population evolve. The evolution process is composed by crossover and mutation. The crossover is applied to the two best individuals (those with highest *AP*) and their genes (states) are crossed. For example, in an 3 state model, state 1 is chosen from individual A and states 2 and 3 are chosen from individual B, generating two new individuals. This stage can be seen in Fig. 6.

$$AP = \frac{\sum P_t}{Q} \tag{4}$$

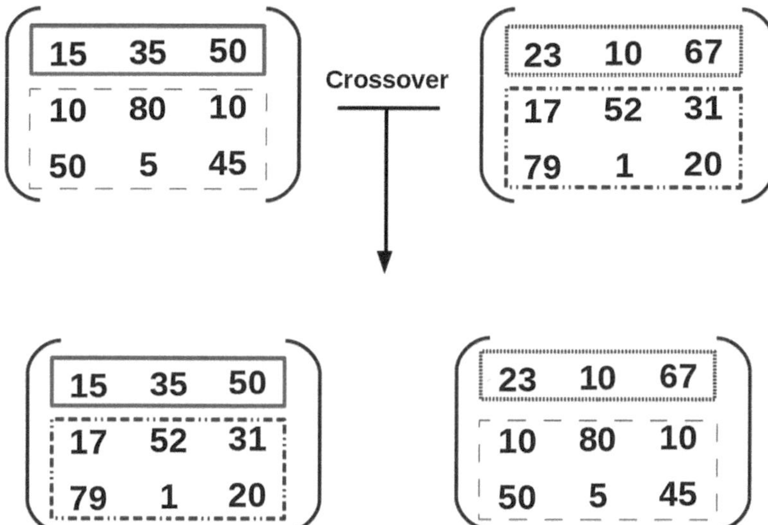

Fig. 6 GMLA crossover process generate two new individuals

After the crossover, all individuals that have an *AP* value below a certain threshold are mutated. The mutation is applied in a state randomly chosen. Transitions are also randomly chosen by applying algorithm 1, so new probabilities are obtained. Is shown in Fig. 7 the result of mutation operation. Therefore, the worst two individuals are replaced by new individuals. These steps are known as the learning phase. They are repeated until an individual with *AP* above the threshold is obtained. Finally, the best individual is chosen (the one with the highest *AP*).

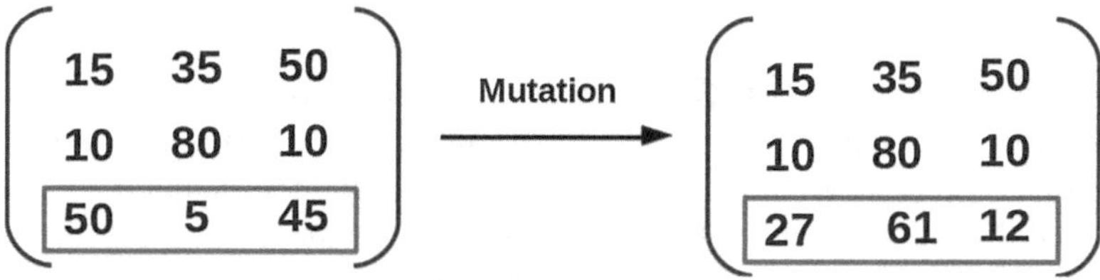

Fig. 7 GMLA mutation process randomly chose the state and apply Algorithm 1

Input: StateTrasitions,numbertransitions
$N = numbertransitions;$
foreach *StateTrasitions i* **do**
  if $i == N$ **then**
    $P_i = 100 - (\sum_{i=0}^{N-1} P_i);$
  end
  else
    $P_i = random(0, 100 - \sum_{i=0}^{j-1});$
  end
end

Algorithm 1: Random selection of transitions probabilities.

6 GMLA Evaluation

This section presents an evaluation of the proposed GMLA. The main goal of this evaluation is to quantify the prediction accuracy of the proposed model and to compare it with existing solutions. The prediction accuracy represents the ability of the model to correctly forecast the next link quality state. In fact, it does not predict the next link quality state just for the next point in time, but for after n seconds ahead (recall that in the model it is assumed at most one state transition per second). The prediction accuracy was expressed here as a metric called rate of accuracy. In the evaluation presented here SNR is used as link quality metric. For a matter of comparison, the original Birth-Death (BD) model and the MTCP [24] were also implemented in our study. Moreover, the deterministic method (Eq. 1) proposed by [25] was also implemented, in order to observe how a deterministic approach behave in three different mobilities patterns which present randomness. We call the last method as GPS.

The evaluation was performed by means of simulations using OMNET++ [28] with the INETMANET [16]. OMNET++ is an oriented-object and event-driven simulation tool. It shares most similarities from NS-2 [26] however, according to [17], OMNET is more friendly for developing the components, since its architecture is modular. As NS-2 and OMNET++ have the same purpose, we decided to use OMNET++, because of its flexibility for developing simulations. In the designed simulations, the wireless nodes communicate by means of the IEEE 802.15.4 standard, using as MAC protocol the non-beacon mode. The GMLA model was implemented here on top of the MAC protocol (network layer).

To perform the simulations it was observed the need for using different mobility models. Therefore we deployed the nodes to behavior as a mobile robots collaborate with each other and also with fixed sensor nodes. As discussed in [5], the mobility models used represent a realistic behavior. The first adopted mobility model is the *Gauss-Markov*, with α parameter randomly selected between 0.5 and 0.9 to avoid complete random and linear moviments. The second adopted is *Reference Point*

Group Mobility Model (RPGM) to represent robots moving in a coordinated manner. The last adopted mobility model is the *Manhattan Grid Mobility Model* (MGM) which was mainly proposed to represent the movement in an urban area.

The reason for choosing a simulation tool to perform this evaluation comes from its flexibility to deal with different mobility models. Since our purpose is to capture the variation of link quality due to mobility, then it is more suitable to use a simulation tool. Therefore, it is possible to observe link quality behaviour in different types of mobility. On the other hand, since it is not possible to use the adopted simulation tool to compute the execution time of the proposed solution, it was also implemented and tested using real sensor nodes, such as the Atmel Atmega128RFA1 [4], which operates at 16 MHz and has 8 KB of SRAM and 128 KB of flash memory the code uploaded in the node occupied 32 KB. Results show that the computing power available on such simple nodes was more than enough to host the execution of the proposed solution. This lead us to the conclusion that the proposed approach can be feasibly implemented on any constrained, simple, computing device (such as a wireless sensor node).

As already mentioned, our approach needs to learn the transition matrix for each individual scenario. In the simulations performed, it was needed the 500 initial seconds of GMLA simulation to complete the so-called learning phase. The other approaches used in the comparison needed twice this time.

6.1 Setup Configuration

Each experiment included 100 nodes sending messages using a random interval between 1 and 5 s. 10 % of the nodes are static (e.g. fixed sensors spread in the factory floor) and the other 90 % of the nodes are mobile (e.g. robots that move around the factory), following one of the three different mobility models presented in the previous section. Additionally, 10 % of the mobile nodes were randomly chosen to be responsible for predicting the connectivity between their neighbors (an uniform distribution was used).

The prediction accuracy of the models was evaluated considering n steps (seconds) ahead. More specifically, for each mobility pattern we have conducted experiments using 1, 5, 10, 15, 20, 25, and 30 steps. Additionally, in order to obtain a better understanding about the influence of the number of states, three different versions were implemented, containing 6, 8, and 10 states. The original BD model was implemented using 8 states, as this number was suggested in [15] as the one that best represents the link quality variation.

The general simulation parameters and the GMLA parameters are presented in Table 1. The number of individuals for GMLA parameters was chosen to cope with the restrictions of a typical sensor node.

Table 1 General parameters for OMNET

Parameters	Value
Sensitivity	-85 dBm
Default transmission power	1.0 mW
Thermal noise	-110 dBm
Playground size	$1000 \times 1000\,\mathrm{m}^2$
Simulation time	20.000 seconds
speed	1 to 10 mps
Attenuation model	Path loss reception model
Number of individuals	16
Mutation threshold	70 %
Number of consults	100

6.2 Obtained Results

The results of the comparison between the three different GMLA models (6,8 an 10 states), the BD, MTCP and GPS models are presented in Fig. 8a–c. The models were compared by means of their prediction accuracies, considering n steps (seconds) ahead. It must also be highlighted that the intention of the experiments was not just to evaluate the models in a quantitative basis. Indeed, it intended to observe the feasibility of using link quality as a way to predict connectivity between mobile nodes in a WSN. Note that this is a very different approach if compared to what is under use in MANETs, as discussed in Sect. 1.

In all experiments GMLA(n) overcomes BD, MTCP and GPS in terms of prediction accuracy. The measures between the different GMLA versions were slightly different. It confirms the fact that as the number of states grows, it gets more difficult to predict. The results also show that GMLA ability (for 6, 8, and 10 states) to foresee the correct next state j from current state i has a respective accuracy average of 98.5, 98.0, and 96.7 % for 1 step ahead in the Manhattan mobility model (best case). For a 30 seconds ahead prediction running the Gauss-Markov model (worst case) it achieves 80, 79, and 77 %. This means that overall the model is able to forecast correctly the next states and, even with 30 s ahead, the result was satisfactory.

The results might also be analyzed by two other perspectives: (i) the influence of the number of states and (ii) the influence of the mobility models. In the first aspect it was noticed that as the number of states grows the accuracy decreases. This is due to fact that more states means more options to choose, suggesting a finer grain prediction. As the measurements itself are not so accurate, it is better to stay with a more general view (smaller number of states). However, there must be a compromise in the protocol design, because if the number of states is too low there might be not enough precision for the protocols to behave properly.

Regarding the mobility pattern, despite the differences in the simulation results we noticed that for Manhattan and RPGM mobility models the results for each GMLA version was quite similar. However, for the Gauss-Markov model, the prediction sharps fast after five steps ahead. The randomness characteristic of this model cre-

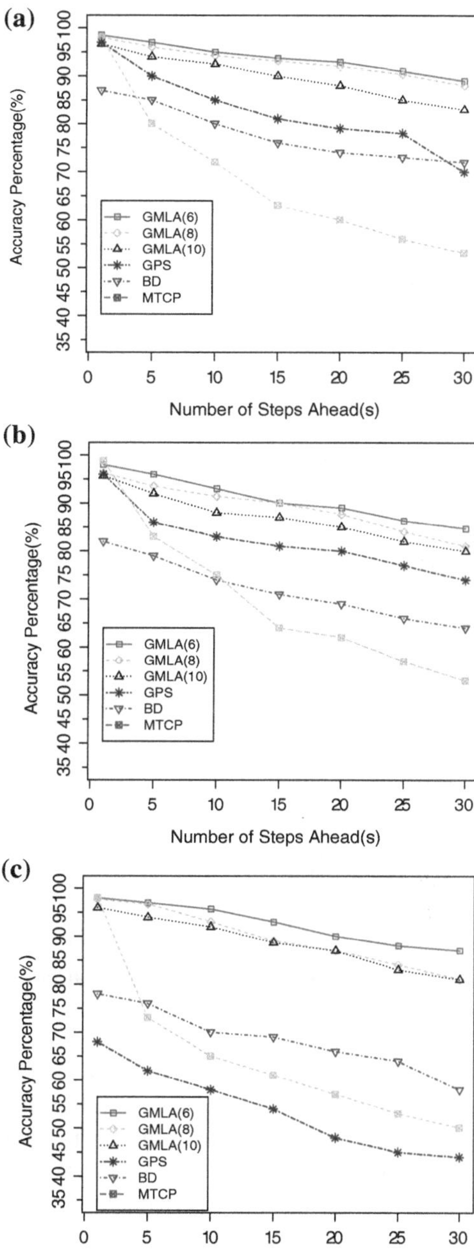

Fig. 8 Accuracy to choose the correct next state in 1, 5, 10, 15, 20, 25 and 30 s ahead for (**a**) Accuracy Percentage For Manhattan Mobility Model, (**b**) Accuracy Percentage For Reference Point Group Mobility Model and (**c**) Accuracy Percentage For Gaussmarkov Mobility Model

ates more difficulty to create a mobility pattern, therefore leading to estimation mistakes. However, the model could work very well with the results of 98 % for GMLA (6, 8) and 96 % from GMLA (10) (in Manhattan mobility) and 80, 79, and 77 % of accuracy in the worst case (Gauss-Markov). For the DB model, the results ranged between 87 % for Manhattan mobility (1 step ahead) and 58 % for Gauss-Markov (30 steps ahead) in the best and worst case respectively. The MTCP model had a good result, about 98 %, for 1 step ahead in all mobility models. However, the results drop fast after 1 step ahead reaching 50 % in worst case for 30 steps ahead in Gauss-Markov Mobility model. The reason for such decrease is that the MTCP is almost a fully connected automaton. It means that each state has more outcome transitions probabilities, then it has more possible final states destinations. The more probability to choose it has, the more difficult is to hit the correct state. The GPS model has interesting results, it had good results for Manhattan and Group mobilities models, ranged between 97 and 70 % for the former mobility and 96 and 74 % for latter mobility. However, for GaussMarkov mobility, the worst results ranged between 68 and 44%. The GPS results point us the limitation of using a deterministic method to predict connectivity in a mobility with the presence of randomness. All mobility models has a randomness behavior, but it is more evident in GaussMarkov mobility. Then, all parameters needed for GPS method could change after the calculation, which lead us to believe that the method must be calculated each time the parameters change.

Finally, the results also show that it is possible to use the knowledge of link quality within ranges to predict the connectivity with a suitable precision. Moreover, it is possible to learn the mobility pattern using the link quality behavior to build the knowledge of connectivity.

7 Conclusions

This work was motivated by the need to improve existing solutions that deal with connectivity prediction, especially when applied to unknown environments and mobility patterns. Existent solutions rely either in a previous history or on some kind of positioning information.

To overcome this problem it was presented the Genetic Machine Learning Approach for Link-quality Prediction (GMLA), which is a solution to forecast the remainder RF connectivity time in mobile environments. Basically, it consists in combining a Classifier System with a Markov chain model of the RF link quality. Its main advantage in comparison to the related works is that it allows the Markov model parameters to be discovered on-the-fly, making it suitable to be used at unknown environments and mobility patterns.

The experiments performed show that GMLA presents a considerable prediction improvement in comparison to existent solutions in different mobility scenarios. Instead of BD and MTCP approaches were designed to predict the link quality behavior, they cannot represent the mobility patterns expressed in link quality behavior.

However, even with the lack of orientation, the DB model is more suitable than the MTCP to represent the link quality variation due to the movement.

Furthermore, the GPS approach only suits well in more deterministic mobility patterns—with less randomness. As the randomness increases in the movement, all parameters needed for GPS solution must be updated and the connectivity estimation must be recalculated.

GMLA can be combined with some existing network protocol either reactive such as AODV (Ad hoc On-Demand Distance Vector) [21], or proactive such as OLSR (Optimized Link State Routing Protocol) [8]. These protocols adopt some messaging mechanism to have connectivity awareness, which can be improved by the GMLA mechanism.

Acknowledgments Thanks are given to the Brazilian research agency CAPES (Coordination for the Improvement of Higher Education Personnel) for its financial contribution under grants 0155-11-0 and 0616-11-7.

References

1. Ali, A., Latiff, L.A., Fisal, N.: GPS-free indoor location tracking in mobile ad hoc network (MANET) using RSSI. In: Proceeding of IEEE RFM, pp. 251–255 (2005)
2. Araújo, G.M.d., Becker, L.B.: A network conditions aware geographical forwarding protocol for real-time applications in mobile wireless sensor networks. In: Proceeding of IEEE AINA. IEEE Computer Soceity, pp. 38–45 (2011)
3. Araújo, G.M.d., Kaiser, J., Becker, L.B.: An optimized Markov model to predict link quality in mobile wireless sensor networks. In: Proceeding of IEEE ISCC. IEEE Computer Society, California, pp. 307–312 (2012)
4. Atmel Atmega128RFA1. http://www.atmel.com/devices/atmega128rfa1.aspx
5. Camp, T., Boleng, J., Davies, V.: A survey of mobility models for ad hoc network research. Wireless communications and mobile computing. Wiley Online Libr. **2**, 483–502 (2002)
6. Chella, A., Lo, G.R., Macaluso, I., Ortolani, M., Peri, D.: Multi-robot Interacting Through Wireless Sensor Networks. Infrastructure, vol. 4733 , pp. 789–796. Springer, Berlin (2007)
7. Chen, S., Jones, H., Jayalath, D.: Effective link operation duration: a new routing metric for mobile Ad hoc networks. In: International Conference on Signal Processing and Communication Systems, Citeseer (2007)
8. Clausen, T., Jacquet, P.: Optimized link state routing protocol (OLSR). RFC 3626, IETF Network Working, Group, Oct 2003
9. Deak, G., Curran, K., Condell, J.: Filters for RSSI-based measurements in a device-free passive localisation scenario. Int. J. Image Process. Commun. **15**, 23–34 (2011)
10. Erman, A.T., Van Hoesel, L., Havinga, P., Wu, J.: Enabling mobility in heterogeneous wireless sensor networks cooperating with UAVs for mission-critical management. IEEE Wireless Commun. **15**, 38–46 (2008)
11. Erman, A.T., Van Hoesel, L., Havinga, P., Wu, J.: Mobile wireless sensor network: Architecture and enabling technologies for ubiquitous computing. Proc. IEEE AINAW **2**, 113–120 (2007)
12. Farkas, K., Hossmann, T., Legendre, F., Plattner, B., Das. S.K.: Link quality prediction in mesh networks. Comput. Commun. **31**, 1497–1512 (2008) (Elsevier)
13. Freitas, E.P.d., Heimfarth, T., Schmidt, R., Wagner, F.R., Larsson, T., Pereira, C.E., Ferreira, A.M.: Coordinating aerial robots and unattended ground sensors for intelligent surveillance systems. Int. J. Comput. Commun. Control Univ. Oradea **5**, 52–70 (2010)

14. Goldberg, D.E.: Genetic algorithms in search, optimization, and machine learning. Addison-wesley, Reading (1989)
15. Guha, R.K., Sarkar, S.: Characterizing temporal SNR variation in 802.11 networks. IEEE Trans. Veh. Technol. **57**, 2002–2013 (2008)
16. INETMANET Framework for OMNEST/OMNeT++ 4.x. http://wiki.github.com/inetmanet/inetmanet/
17. Koksal, M. M.: A survey of network simulators supporting wireless networks, Middle East Technical University Ankara, TURKEY, 22 Oct 2008
18. Lee, S.J., Su, W., Gerla, M.: Mobility prediction in wireless networks. In: Proceeding of IEEE ICCCN 2000, Boston, MA, p. 49 (2000)
19. Liu, T., Sadler, C.M., Zhang, P., Martonosi, M.: Implementing software on resource-constrained mobile sensors: experiences with Impala and ZebraNet. Proc MobiSys, pp. 256–269. ACM, New York (2004)
20. Nicholson, A.J., Noble, B.D.: Breadcrumbs: forecasting mobile connectivity. In: Proceeding of ACM MobiCom, pp. 46–57 (2088)
21. Perkins, C., Belding-Royer, E., Das, S.: Ad hoc on-demand distance vector (AODV) routing. RFC 3561, IETF Network Working Group, July 2003
22. Priyantha, N.B., Miu, A.K., Balakrishnan, H., Teller, S.: The cricket compass for context-aware mobile applications. In: Proceeding of ACM MobiCom, pp. 1–14 (2001)
23. Rosa, F.d., Malizia, A., Mecella, M.: Disconnection prediction in mobile ad hoc networks for supporting cooperative work. IEEE Pervasive Comput. **3**, 62–70 (2005)
24. Sabitha, R., Thangavelu, T.: Performance enhancement of fuzzy logic based transmission power control in wireless sensor networks using Markov based RSSI prediction. Eu. J. Sci. Res. Euro J. Pub. **59**, pp. 68–84 (2011)
25. Su, W., Lee, S., Gerla, M.: Mobility prediction in wireless networks. In: Proceeding of IEEE ICCCN. IEEE, New York, pp. 4–9 (1999)
26. The Network Simulator - ns-2. http://www.isi.edu/nsnam/ns/
27. Valente, J., Sanz, D., Barrientos, A., Cerro, J., Ribeiro, Á., Rossi, C.: An Air-Ground Wireless Sensor Network for Crop Monitoring. Sensors **11**, 6088–6108 (2011)
28. Varga, A.: The OMNeT++ discrete event simulation system. In: Proceeding of ESM, pp. 319–324 (2001)

Chapter 2
Generation of Trajectories Using Predictive Control for Tracking Consensus with Sensing and Connectivity Constraint

Bernardo Ordoñez, Ubirajara F. Moreno, Jés Cerqueira and Luis Almeida

Abstract This work presents a cooperation strategy for teams of multiple autonomous vehicles to solve the rendezvous problem. The approach is based on consensus algorithms, which are basically characterized by information exchange among the team members. The proposal is based on predictive control in order to compute decentralized control laws, considering constraints and different response requirements according to the application scenario, for example, constraints related to coverage and connectivity of the group. Our work allows considering together vehicles without and with non-holonomic restrictions while optimizing the sensing range, particularly that of fixed frontal cameras, by managing orientation in the way to the rendezvous point. We show the effectiveness of our strategy with simulation results.

Keywords Consensus algorithm · Cooperation strategies · Optimization · Non holonomic constraint

This work was supported in part by CAPES/Brazil (Coordenação de Aperfeiçoamento de Pessoal de Nível Superior) and FCT/Portugal (Fundação para Ciência e Tecnologia).

B. Ordoñez (✉) · U. F. Moreno
Department of Automation and Systems Engineering, Federal University of Santa Catarina, BOX 88040-900, Florianópolis, Santa Catarina, Brazil
e-mail: ordonez@das.ufsc.br

U. F. Moreno
e-mail: moreno@das.ufsc.br

J. Cerqueira
Department of Electrical Engineering, Federal University of Bahia, BOX 40210-630, Salvador, Bahia, Brazil
e-mail: jes@das.ufsc.br

L. Almedia
IT—Faculty of Engineering, University of Porto, BOX 4200-465, Porto, Portugal
e-mail: lda@das.ufsc.br

A. Koubâa and A. Khelil (eds.), *Cooperative Robots and Sensor Networks*,
Studies in Computational Intelligence 507, DOI: 10.1007/978-3-642-39301-3_2,

1 Introduction

The use of robotic vehicles to perform tasks autonomously is becoming widespread due to both technological and scientific advances, for example, the miniaturization of electromechanical systems and new sensing and control paradigms. It is natural to imagine that soon, teams of vehicles will be fully autonomous and capable of carrying out challenging tasks. The use of autonomous vehicles requires coordination through the use of cooperation strategies because there are tasks that one vehicle alone could not perform due to both its partial knowledge about the task and limited resources. A coordinated set of vehicles can share information in dynamic environments to perform challenging tasks [1, 2]. Examples can be found in applications that range from military systems to mobile surveillance sensor networks for monitoring roads and air transport systems [3].

The concept of cooperative control implies by definition that in a task performed by a team of vehicles these can communicate and collaborate [4]. Among the main techniques used to solve tasks in a cooperative way, this paper focuses on consensus algorithms, which are characterized by communication and information exchange within the team [5]. Another important feature of this technique is that the design of consensus algorithms is based on decentralized implementation.

The main studies about consensus were focused on the algorithm analysis while requirements such as control effort and tracking error were not considered. Therefore, this work presents a technique for synthesis of decentralized control laws to generate consensus trajectories that maximize the performance with respect to response requirements. The strategy was named by ACvO (Algorithm Consensus via Optimization). The contributions of our strategy are twofold, on one hand, non-holonomic constraints of vehicles motion are considered when defining the tracking trajectories, on the other hand, we optimize the sensing range, particularly of fixed front cameras, according to the rendezvous point, in the sense that we control the orientation at which the vehicles arrive at the rendezvous.

A preliminary version of this work was presented in [6], in which the main result was the development of feasible consensus trajectories for all vehicles, including non-holonomic constraints and sensing range optimization. In the current work we add a discussion on the convergence of consensus trajectories and the computational time for each algorithm iteration. Moreover, we show the advantage of using the method to add constraints in a straightforward way, and we illustrate it adding an important group connectivity constraint to achieve global consensus.

The paper is organized as follows. Section 1 presented the problem definition and the main related works about cooperative control and consensus algorithm. The aspects and definitions about the consensus algorithms are shown in Sect. 2. Section 3 presents the proposed formulation via optimization to perform consensus trajectories for the rendezvous problem. Section 4 presents numerical simulations in order to analyze the proposed strategy. Also it is introduced a discussion about the convergence and computational time of the ACvO protocol. Finally, Sect. 5 concludes the work with a discussion of results and future work.

1.1 Related Works

The consensus problem has a long history related to the field of computer science. It was one of the bases for the development of distributed computing. A historical perspective about consensus algorithms can be founded in [7] and [8], where the study of consensus problem was formalized. Although there is this historical relationship, the focus of this work is on the application of consensus theory to cooperative control for multi-vehicles group coordination.

Theoretical aspects about the definition and resolution of the consensus problem were introduced in [9] and [10]. In these works, the consensus problem in vehicles dynamic networks is analyzed for three cases: directed networks with fixed topology, directed networks with switching topology, and network topology and communication with fixed delay. The characteristic that distinguishes those works is the consensus approach based on networks with directed information flow. The work in [11] addresses the problem of multi-vehicles cooperation with the application of algorithms based on graph theory that relates the network communication topology to the vehicles formation stability.

The consensus problem with multi-vehicles group in the presence of limited and uncertain information flow due to time-varying topologies is considered in [12]. In that work the important concept of directed spanning tree is used to evaluate the consensus algorithm convergence and graph connectivity.

A typical application of the consensus algorithms in the cooperative control context is the *rendezvous problem*, which is characterized by a group of vehicles that negotiate among them to determine a meeting location or time. Thus, the rendezvous problem for multi-vehicle groups can be defined as the design of local control strategies without active communication between the vehicles to determine the meeting point.

The rendezvous problem is also present in air surveillance tasks, where the use of UAVs has grown considerably. For example, the work in [13] shows an approach to the cooperation problem that uses a consensus algorithm to replace a centralized strategy with a decentralized algorithm. The validation was carried out using numerical simulations.

All the works related to cooperative control algorithms and consensus mentioned above addressed the possibility of adjustments in the algorithms and strategies to improve performance of a particular requirement, such as the convergence rate to reach consensus, minimization of the tracking error, or even the vehicles control effort.

The field of optimal control algorithms has been extensively studied, especially in this last decade. The development of optimal coordination strategies based on consensus algorithms is seriously compromised by the presence of corrupted data and uncertainties in measurements. Therefore, some techniques were developed to ensure cooperation among vehicles with limited access to information [9]. In [14] the convergence rate for a classes of consensus is studied using semi-defined programming. In [15], necessary and sufficient conditions of convexity related to system topology are developed.

In [16] the strategy of receding horizon control is used to formation stabilization with quadratic cost and no coupling constraints. The consensus problem in a group of vehicles with time-varying topologies is studied in [17], in which the vehicles only need knowledge about the neighboring states to reach consensus. A similar approach is found in [18], where the objective is to develop controllers for a group of vehicles based on means of consensus. An optimal control semi-decentralized strategy is developed based on the minimization of individual cost functions with finite horizon and local information.

A formation control law based on artificial potential fields and consensus algorithms for a group of unicycles is proposed in [19], which considers connected and balanced graphs to prove stability of the controller by applying the LaSalle-Krasovskii invariance principle. The work in [20] addresses the design of optimal control to ensure consensus in a multi-vehicle network adopting a global cost function to ensuring consensus with optimal control effort. It is shown that the solution of Riccati equation does not guarantee consensus under certain conditions, and therefore, it proposes a formulation based on LMI to resolve the optimization problem with constraints.

Finally, [21] addresses the parameters weight in a problem of convex optimization for random topologies, in which the convergence rate square error is used as optimization criterion. The work in [22] studies optimal consensus algorithms from the perspective of LQR theory, in which the Laplacian matrices have direct influence on the choice of optimal parameters of the system, while [23] shows a distributed cooperation algorithm for problems with coupling hard state constraints (non-convex and external disturbances).

2 Consensus Algorithms

In the context of cooperative control, *consensus* can be defined by a commitment among the group members to a common goal (group decision value). A variable defined as *information state* is used to model the collective view of the common objective and it can be used to represent some abstractions of the coordination variable, such as rendezvous location.

Let a directed graph of order n be represented by $G_n = (\nu_n, \varepsilon_n)$ with the set of nodes $\nu = \{v_1, ..., v_n\}$, set of edges $\varepsilon_n \subseteq \nu_n \times \nu_n$ and n being the number of vehicles. The nodes belong to a finite index set $\Gamma = 1, ..., n$. An edge of G_n is denoted by $e_{ij} = (v_i, v_j)$. The adjacency elements (a_{ij}) associated with the edges of the graph are positive, i.e., $e_{ij} \in \varepsilon \Leftrightarrow a_{ij} > 0$. It is assumed that $a_{ii} = 0$ for all $i \in \Gamma$. Finally, the set of neighbors of node ν_i is denoted by $N_i = v_j \in \nu : (v_i, v_j) \in \varepsilon$.

Definition 1 *(Directed Tree)* A directed tree is a digraph G_n, where exists a vehicle named root, such that all of other vehicles of the digraph can be reached through one path only starting at the root. Consequently, $\tau_G = \{\nu_\tau, \varepsilon_\tau\}$ is a spanning tree of G_n, if both τ_G is a directed tree and $\nu_\tau = \nu$.

Let $\xi_i \in \Re^n$ denote the decision group value of node v_i, then, $G_\xi = (G_n, \xi)$ with $\xi = (\xi_1, ..., \xi_n)^T$ representing a network with communication topology (or information flow) G_n. Suppose each node v_i of the digraph G_ξ has the following dynamics where $u_i \in \Re^n$ is the input control signal of the ith vehicle:

$$\dot{\xi}_i = f(\xi_i, u_i), \qquad i \in \Gamma \tag{1}$$

then, we can define a digraph as a dynamic system represented by $G_\xi = (G_n, \xi)$, in which the evolution ξ_i is governed by the network dynamics $\dot{\xi}_i = f(\xi_i, u_i)$. Let the information state with single integrator dynamic be given by:

$$\dot{\xi}_i(t) = u_i(t), \qquad i = 1, ..., n. \tag{2}$$

The basic consensus protocol can be defined by:

$$u_i(t) = -\sum_{j=1}^{n} a_{ij}(t)\left(\xi_i(t) - \xi_j(t)\right), \qquad i = 1, ..., n \tag{3}$$

where $a_{ij}(t)$ is the input of adjacency matrix $A_n \in \Re^{n\times n}$ associated to $G_\xi(t)$ and related to the vehicle i and its neighbor j. Note that $a_{ij}(t) > 0$ when $(i, j) \in \varepsilon_n$, otherwise, $a_{ij}(t) = 0$.

Definition 2 *(Average Consensus)* The consensus for the multi-vehicle network (2) is achieved when for every initial state $\xi_i(0)$,

$$lim_{t\to\infty} \left|\xi_i(t) - \xi_j(t)\right| = 0,$$

for $i = 1, ..., n$ and $j = 1, ..., n$.

The dynamic of the *information states* (2) can be implemented using the discrete model given by:

$$\xi_i[k+1] = \xi_i[k] + \Delta_k u_i[k], \qquad i = 1, ..., n \tag{4}$$

where Δ_k is the step size, and $\xi_i[k]$ and $u_i[k]$ are the information state and control input of the ith vehicle at discrete time k.

2.1 Consensus Tracking Protocol

The consensus tracking brings the information states of all vehicles to a reference state. Note that from Eq. (3), the consensus equilibrium is a weighted average of all vehicles initial states and hence constant. The consensus value is related to the interaction topology and weights a_{ij} of the adjacency matrix and it is unknown *a priori*.

However, in some applications it is desirable that the consensus information states converge to a predefined value. In these cases, the convergence issues include both convergence to a common value, as well as convergence of the common state to its reference value. Therefore, let us consider a group with n vehicles plus an additional and virtual leader $n+1$. The state $\xi_{n+1} = \xi_r \in \Re^n$ contains the information about reference consensus.

Definition 3 *(Tracking Consensus)* The tracking consensus in the multi-vehicle network, (2), is achieved when for every initial state $\xi_i(0)$,

$$lim_{t\rightarrow\infty} \left|\xi_i(t) - \xi_j(t)\right| = 0$$

and

$$lim_{t\rightarrow\infty} |\xi_i(t) - \xi_r(t)| = 0$$

for $i = 1, ..., n$ and $j = 1, ..., n$.

The digraph $G_{n+1} = (\nu_{n+1}, \varepsilon_{n+1})$ is used to model the interaction among the $n+1$ vehicles (with a virtual leader). Let $A_{n+1} = [a_{ij}] \in \Re^{n+1\times n+1}$ the adjacency matrix associated to G_{n+1}, where $a_{ij} > 0$ if $(j, i) \in \varepsilon_{n+1}$ and $a_{i(n+1)} > 0$ if ξ_r is available to vehicle i for $i = 1, ..., n$. Finally, $a_{(n+1)j} = 0$ for all $j = 1, ..., n+1$ and $a_{ii} = 0$ for all $i = 1, ..., n$.

From [4], we have the following theorem for consensus tracking with a constant consensus reference state.

Theorem 1 *Suppose that A_{n+1} is constant. The consensus tracking problem with a constant consensus reference state is solved with according to Definition 3 if and only if the directed graph G_{n+1} has a directed spanning tree.*

Note that vehicle $n+1$ has the information about the reference and the condition that G_{n+1} has a directed spanning tree is equivalent to the condition that, at least, the virtual vehicle $n+1$ has a directed path to all other vehicles in the team, which is a guarantee to the tracking consensus of the *information states*.

The assumption that the matrix A_{n+1} is constant is guaranteed by the connectivity constraint added to the ACvO in this work. We leave the case of varying topologies (non-constant A_{n+1}) for future work in which we will explore the constraints on the variability of A_{n+1} that still allow reaching consensus.

Assumption 1 *We assume that the digraph $G_{n+1}(t|t=0)$ has a directed spanning tree.*

The *Assumption 1* guarantees that the initial oriented communication graph, $G_{n+1}(t|t=0)$, has, at least, one path connecting all the vehicles, including the virtual leader, $n+1$, which, contains the reference information.

The main goal of the next section is to develop control laws that guarantee that each vehicle of the group achieves trajectory consensus, which is only known by a subset of the group.

3 Synthesis of Control Laws via Optimization for Consensus Trajectories

This section formulates a methodology for control laws synthesis based on consensus theory as an optimization problem, and therefore, it was named by ACvO (Algorithm Consensus via Optimization). As previously mentioned, the main studies found in the literature about vehicle consensus approaches were focused on the algorithm analysis, and consequently the synthesis of control laws is ignored. Therefore, requirements such as control effort, control signal saturation, convergence rate and tracking error were not considered in such previous work. The first challenge is to define an objective function, J_i, with the commitment between performance indexes of requirements response and cooperation terms, especially regarding the information exchange among vehicles.

Let the preliminary J_i function be as follows:

$$J_i[k] = \sum_{j=1}^{n}\sum_{k=1}^{N_p}\left(\hat{\xi}_i[k]-\hat{\xi}_j[k]\right)'\delta_\xi\left(\hat{\xi}_i[k]-\hat{\xi}_j[k]\right) + \sum_{k=1}^{N_p}\left(\hat{\xi}_i[k]-\xi_r[k]\right)'\delta_e\left(\hat{\xi}_i[k]-\xi_r[k]\right), \qquad (5)$$

where n is the number of vehicles, N_p is the horizon of prediction and $\hat{\xi}_i$ is prediction of state ξ_i, for $i=1, ..., n$ with J_i corresponding to the ith vehicle.

The matrices δ_ξ and δ_e are composed according to the values of the adjacency matrix. When there is no channel of communication between i and j vehicles, the input parameters of the matrix δ_ξ are zero. In similar way, when the vehicle i has no information about the reference, the input parameter of the matrix δ_e is zero.

The objective function presented in Eq. (5) is familiar to control laws widely utilized in the literature [1, 4, 24, 25]. Both control laws and objective function represent the trade-off between states energy and tracking reference. The advantage of the proposed methodology is to precisely include requirements that are not yet included in the optimization problem. The inclusion of control effort is straightforward, just adding:

$$J_i^{aux}[k] = \sum_{k=1}^{N_u} (\Delta u_i[k])' \lambda_u[k](\Delta u_i[k]), \tag{6}$$

where, N_u is the control horizon, $\Delta u_i[k]$ is the control increment and $\lambda_u[k]$ is a math function that represents the future behavior of the system. An objective function composed by Eqs. (5) and (6) includes implicitly a trade-off among control effort, states energy and tracking error.

The objective functions (5) and (6) in the matrix form are given by:

$$\begin{aligned} J_i[k] = &\sum_{j=1}^{n} \left(\hat{E}_{\xi_i}[k] - \hat{E}_{\xi_j}[k]\right)' \Delta_{\xi_i} \left(\hat{E}_{\xi_i}[k] - \hat{E}_{\xi_j}[k]\right) \\ &+ \left(\hat{E}_{\xi_i}[k] - E_{\xi_r}[k]\right)' \Delta_{e_i} \left(\hat{E}_{\xi_i}[k] - E_{\xi_r}[k]\right) \\ &+ \left(\Delta u_{\xi_i}[k]\right)' \lambda_{u_i} \left(\Delta u_{\xi_i}[k]\right), \quad \text{for } i = 1, ..., n, \end{aligned} \tag{7}$$

where,

$$\begin{aligned} \hat{E}_{\xi_i}[k] &= \left[\xi_i[k+1|k]' \;\; \xi_i[k+2|k]' \;\; \ldots \xi_i[k+Np|k]'\right]' \\ \hat{E}_{\xi_j}[k] &= \left[\xi_j[k+1|k]' \;\; \xi_j[k+2|k]' \;\; \ldots \xi_j[k+Np|k]'\right]' \\ E_{\xi_r}[k] &= \left[\xi_r[k+1|k]' \;\; \xi_r[k+2|k]' \;\; \ldots \xi_r[k+Np|k]'\right]' \\ \Delta u_{\xi_i}[k] &= \left[\Delta u_i[k|k]' \;\; \Delta u_i[k+1|k]' \;\; \ldots \Delta u_i[k+Np-1|k]'\right]'. \end{aligned}$$

The prediction of states to horizon N_p, according to Eq. (4), and the increment of the control inputs can be presented under the following matrix form:

$$\hat{E}_{\xi_i} = E^0_{\xi_i} + T\, U_i \tag{8}$$

$$\Delta u_{\xi_i} = U^0_{\xi_i} + U_{aux} U_i, \quad \text{for } i = 1, ..., n \tag{9}$$

where, $\hat{E}_{\xi_i} \in \Re^{N_p \times 2}$ is a states prediction matrix. $E^0_{\xi_i} \in \Re^{N_p \times 2}$ is a matrix with state ξ_i at begin of the horizon prediction k, $T \in \Re^{N_p \times N_p}$ is a matrix composed by Δ_k and $U_i \in \Re^{N_p \times 2}$ is a vector with the future control inputs. In second line, $\Delta u_{\xi_i} \in \Re^{N_u \times 2}$ is a matrix of control increments, $U^0_{\xi_i} \in \Re^{N_u \times 2}$ is a matrix with control U_{ξ_i} at time k, $U_{aux} \in \Re^{N_u \times N_u}$ is an auxiliary matrix to process the difference between increments of control and $U_i \in \Re^{N_u \times 2}$ is a vector with future control inputs.

Remark 1 Note that Eq. (7) contains both control inputs of vehicles i and j (U_i and U_j, respectively). Remember that since the control law is decentralized, for each vehicle i, the *information state* update can not include the future control inputs of the neighbor vehicle. Note that, to solve the problem for U_i, the optimal future values of U_j are unknown yet, at time k.

A possible solution to the problem stated in *Remark 1* is to rewrite the decision vector as $U = [U_1, ..., U_n]$. However, implementing this arrangement, the problem becomes centralized, and more, it neglects the directed graph characteristic, i.e., input a_{ij} has the same meaning that a_{ji}. Instead we assume that the neighboring prediction states are unknown, since it is not possible to use the optimal control sequence of the neighboring state. The objective function J_i only implements the current state k of the neighbor j. The new objective function is given by:

$$\begin{aligned} J_i^{new} =& \sum_{j=1}^{n} \left(E_{\xi_i}^0 + T\ U_i - E_{\xi_j}^0)' \Delta_{\xi_i} (E_{\xi_i}^0 + T\ U_i - E_{\xi_j}^0\right) \\ &+ \left(E_{\xi_i}^0 + T\ U_i - E_{ref})' \Delta_{e_i} (E_{\xi_i}^0 + T\ U_i - E_{ref}\right) \\ &+ \left(U_{\xi_i}^0 + U_{aux} U_i)' \lambda_{u_i} (U_{\xi_i}^0 + U_{aux} U_i\right) \end{aligned} \tag{10}$$

with $i, j = 1, ..., n$. Separating the terms with U_i in Eq. (10) leads to:

$$\begin{aligned} J_i^{new} =& \sum_{j=1}^{n} U_i' \left(T' \Delta_\xi T + T' \Delta_e T + U_{aux}' \lambda_{u_i} U_{aux}\right) U_i \\ &+ 2 \left(E_{\xi_i}^{0\,'} \Delta_\xi T - \sum_{j=1}^{n} \left(E_{\xi_j}^{0\,'} \Delta_\xi T \right) + E_{\xi_i}^{0\,'} \Delta_e T - E_{ref}' \Delta_\xi T + U_{\xi_i}^{0\,'} \lambda_{u_i} U_{aux} \right) U_i \\ &+ \left(E_{\xi_j}^{0\,'} \Delta_\xi E_{\xi_j}^0 + E_{\xi_i}^{0\,'} \Delta_\xi E_{\xi_i}^0 - 2 (E_{\xi_j}^{0\,'} \Delta_\xi E_{\xi_i}^0 + U_{\xi_i}^{0\,'} \lambda_{u_i} U_{\xi_i}^0) \right). \end{aligned} \tag{11}$$

3.1 Quadratic Formulation Problem

Our goal is to minimize the cost function J_i^{new}, therefore, the constant terms in Eq. (11) can be eliminated. Moreover, defining the auxiliary matrices H_i and f_i as follows:

$$\begin{aligned} H_i &= T' \Delta_\xi T + T' \Delta_e T + U_{aux}' \lambda_{u_i} U_{aux} \\ f_i &= E_{\xi_i}^{0\,'} \Delta_\xi T - \sum_{j=1}^{n} \left(E_{\xi_j}^{0\,'} \Delta_\xi T \right) + E_{\xi_i}^{0\,'} \Delta_e T - E_{ref}' \Delta_\xi T + U_{\xi_i}^{0\,'} \lambda_{u_i} U_{aux}. \end{aligned}$$

allows expressing the minimization of J_i^{new} in Eq. (11) as a quadratic formulation problem given by:

$$\begin{aligned} \min_{U_i} \quad & \frac{1}{2} U_i^T H_i U_i + f_i^T U_i \\ s.t. \quad & c_{\min} \le U_i \le c_{\max} \end{aligned} \tag{12}$$

This methodology allows the addition of constraints into the optimization problem in straightforward way. For example, the control signal saturation is implemented defining minimum and maximum values, $\underline{U} = c_{\min}$ and $\bar{U} = c_{\max}$, respectively, as shown in Eq. (12).

3.2 On the Consensus Trajectories and Vehicles Sensing

The ACvO protocol computes the consensus trajectory associated with the *information state*, only, which is the understanding of each vehicle about the meeting point. We assume the use of local controllers to ensure that the vehicle targets the desired position at every sampling time. It does not consider possible mechanical constraints in the motion of the vehicle, e.g., the orientation constraints of non-holonomic mobile robots.

Figure 1a shows an illustrative example, where a vehicle with local controller, even following the trajectory can fail in meeting non-holonomic constraints (*lower trajectory*). The vehicle orientation, at time k, is opposed to the direction of the next point, and thus, because of non-holonomic constraint, the vehicle makes a turn (*upper trajectory*), delaying its route to the consensus point, needing one more iteration to reach the trajectory.

Moreover, assuming that the vehicle has a fixed camera, in which the sensing range can be associated to the vehicle orientation, we added an optimization routine on the sensing range motivated by the knowledge of the future control sequence (optimization of J_i) and hence, a prediction of all points of the trajectory. The opti-

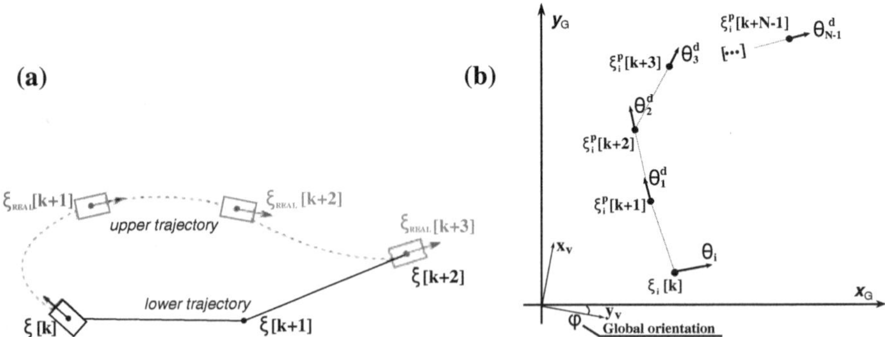

Fig. 1 **a** Illustrative example with non-holonomic constraint. **b** Optimization problem to sensing orientation

mization of vehicle orientation can be performed by minimizing the squared error, according to Fig. 1b.

Let the cost function:

$$J_i^{\theta} = \sum_{k=1}^{N_p-1} \left\| \theta_i^d[k+1] - \theta_i[k+1] \right\|^2, \qquad i = 1, ..., n \tag{13}$$

where, θ_i^d is the desirable orientation (for sensing purposes) and θ_i is the actual orientation of the ith vehicle. The goal is to find the value of θ_i that minimizes J_i^{θ}. Some constraints can be imposed into the problem, such as the maximum individual rotation δ_d and curvature radius r_c of the vehicle. Thus, the new *information states* associated to coordinates x and y are:

$$\begin{aligned} \xi_i^x[k+1] &= \xi_i^x[k] \cos(\theta_i[k+1] + \varphi) \\ \xi_i^y[k+1] &= \xi_i^y[k] \sin(\theta_i[k+1] + \varphi) \end{aligned} \tag{14}$$

where φ is the rotation related to global reference, since δ_d and θ_i are local variables, Fig. 1b. Note that each iteration of the ACvO, the maximum rotation is $3\delta_d$ (due to mechanical constraints and saturation of the control signal) and the final value of θ_i is defined by optimizing J_i^{θ}.

It is important to note that in Eq. (14), the control values do not appear, these values have already been calculated in the previous section. This procedure is complementary, and the goal is to correct the orientation of the vehicle, which may or may not have implication on the computed consensus trajectory.

3.3 On the Connectivity Constraint

As previously mentioned, one of the advantages of the proposed method is the possibility of adding constraints to the problem in straightforward way. In the optimization problem in Eq. (12), it is possible to add inequality constraints ($AU_i \leq B$) on the *information states* according to the application.

For example, conditions may be formulated to avoid collision between the vehicles, but also related to obstacles in the environment, and for UAVs, imposing a minimum height of flight. In this paper we are concerned with one situation in particular, where it is formulated a constraint related to the connectivity of the topology, and it aims at keeping any two neighbor vehicles within a *connectivity radius*.

In the connectivity constraint the main idea is that once defined that i is a neighbor of j, then they will be neighbors throughout the complete task. Thus, if there is a temporary failure in the communication channel, the restriction aims at maintaining the neighborhood relationship between the vehicles. Let $\xi_i[k+1]^{(x,y)} = \xi_i[k] + \Delta_k U_i[k]$ and $\xi_j[k+1]^{(x,y)} = \xi_j[k] + \Delta_k U_j[k]$, the Euclidean distance between the next two positions is:

$$
\begin{aligned}
d_{ij} &= \sqrt{\left(\xi_i^x[k+1] - \xi_j^x[k+1]\right)^2 + \left(\xi_i^y[k+1] - \xi_j^y[k+1]\right)^2} \\
&= \sqrt{\left(\xi_i^x[k] + \Delta_k U_i^x[k] - \xi_j^x[k] + \Delta_k U_j^x[k]\right)^2 + \left(\xi_i^y[k] + \Delta_k U_i^y[k] - \xi_j^y[k] + \Delta_k U_j^y[k]\right)^2}
\end{aligned}
\tag{15}
$$

Again making reference to Remark 1, while the ACvO is calculating the future controls U_i, the values of U_j are still unknown. Then, the new value of the next point of the neighbor j is estimated with a simple derivative. By using the Euclidean distance, the decision variable U_i assumes a quadratic form. To avoid this, we use as distance metric, the *Manhattan form*, which will provide a more conservative result. The following constraint is considered:

$$
\begin{aligned}
d_{ij}^{Mah} &= \left|\xi_i^x[k] + \Delta_k U_i^x[k] - \tilde{\xi}_j^x[k+1]\right| + \left|\xi_i^y[k] + \Delta_k U_i^y[k] - \tilde{\xi}_j^y[k+1]\right| \\
&\leq r_{com},
\end{aligned}
\tag{16}
$$

where, $\tilde{\xi}_j^{(x,y)}[k+1]$ is the estimate state of the neighbor using the derivative term and r_{com} is the radius of communication. This constraint is added in a straightforward way in the optimization problem in the form $AU_i \leq B$.

4 Implementation of Consensus Protocol Strategy

The consensus strategy is implemented based on the blocks diagram shown in Fig. 2a. At each ACvO iteration, the optimization of J_i generates the trajectory to N_p points horizon based on the information exchanged, and more, with the knowledge of all trajectory prediction, the optimal orientation of vehicle ith is also calculated (i.e., optimization of J_i^θ). (Note that the calculus of J_i^θ is only performed to the vehicles that have non-holonomic constraints). However, only the first point is implemented and we assume that the vehicle has a controller and local sensing to achieve this local target point. As a result, we have a feasible consensus trajectory for all vehicles.

The simulations were performed in *Matlab*, where we considered some communication errors (it is admitted temporary fail in some communication channels between vehicles). Note that these losses do not change the communication topology permanently. Since, the failures are temporary, the effect on the network topology just lasts a few algorithm iterations, and then, the channel is reestablished.

The topologies shown in Fig. 2b and c are used in order to evaluate the performance of the proposed algorithm. The topologies were chosen arbitrarily and the arrows indicate the direction of information flow. In Fig. 2, nodes 1 to 6 represent the vehicles while their respective *information states*, are expressed in ξ_n, with $n = 1, ..., 6$. We define that vehicles 1 and 2 are WMRs with non-holonomic constraint, 3 and 4 have no motion restriction and 5 and 6 are UAVs. The state ξ_r is used to define the

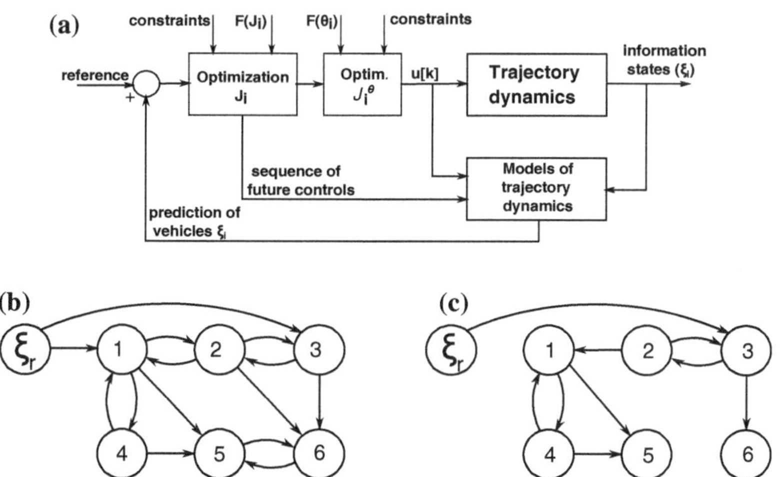

Fig. 2 **a** Blocks diagram to implement consensus strategy via optimization; **b** Topology A; **c** Topology B. Both topologies are used in illustrative examples applying the ACvO algorithm

reference information ($\xi_r^A = [100, 100]$ and $\xi_r^B = [190, 100]$ to topologies A and B, respectively) and its information is available to a few vehicles in the group, only.

In few words, the simulation scenario is characterized by oriented information flow, limited knowledge and communication, and group with heterogeneous vehicles.

Figure 3 shows two different arrangement to the initial positions of all vehicles. The initial positions were defined as follows: in topology A, $\xi_1 = [50, 120]$, $\xi_2 = [120, 180]$, $\xi_3 = [170, 180]$, $\xi_4 = [70, 30]$, $\xi_5 = [40, 50]$ and $\xi_6 = [130, 150]$; and in topology B, $\xi_1 = [35, 100]$, $\xi_2 = [50, 150]$, $\xi_3 = [80, 190]$, $\xi_4 = [45, 30]$, $\xi_5 = [20, 30]$ and $\xi_6 = [40, 195]$.

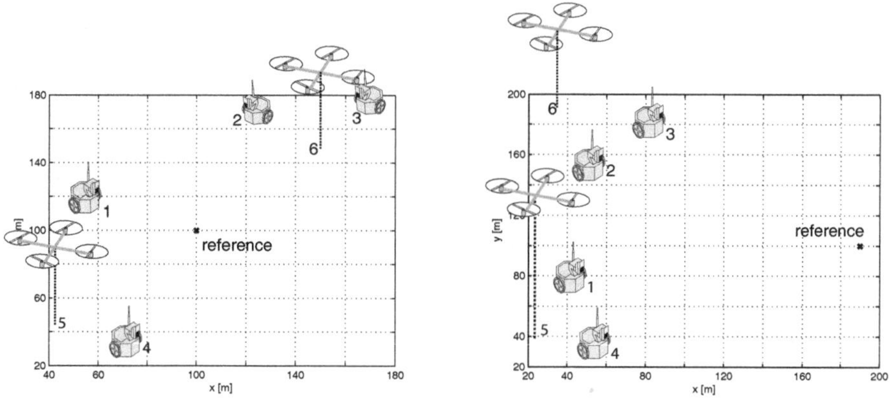

Fig. 3 Initial positions of vehicles for topology A (*left*) and topology B (*right*)

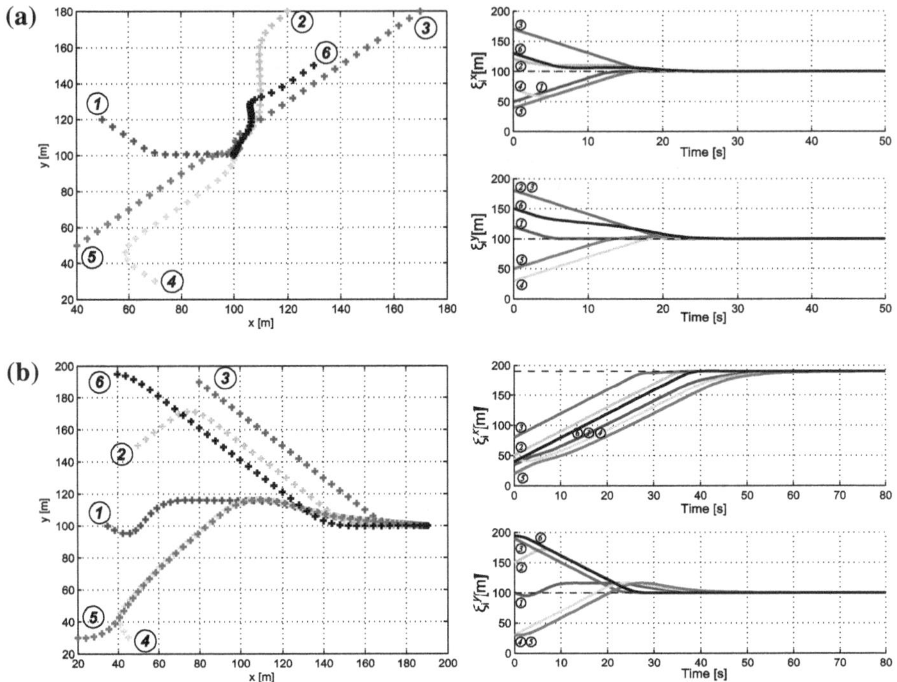

Fig. 4 Dynamics of the *information states* ξ_i applying ACvO protocol without integration of mobile robots sensing. **a** Topology A and **b** topology B

4.1 Numerical Simulation and Analysis of Results

Figure 4 shows the results obtained with the application of the ACvO protocol without the optimization of sensing angle of the vehicles. In other words, only the consensus tracking and control effort are considered in the objective function J_i^{new}, Eq. (11).

The first goal using only the optimization of J_i^{new} is to evaluate the impact of the simplification made in the objective function, in which we only implemented the current state of the neighbors (*Remark 1*). Figure 5 shows the number of iterations required for tracking consensus using a comparison of the decentralized and centralized (knowledge of predicted states in the neighbors) approaches. As a complement, we added a comparison with an algorithm with fixed parameters presented in [4] because it is very similar to other algorithms found in literature [1, 10, 11, 24].

As expected, the results show that the centralized approach has a faster convergence to the tracking consensus, but the difference between the performances is not significant (only ACvO vs. ACvO centralized). In comparison with the algorithm with fixed parameter the results presented reaffirm that the application of the ACvO allows a quick convergence of the vehicles *information state* to the reference state.

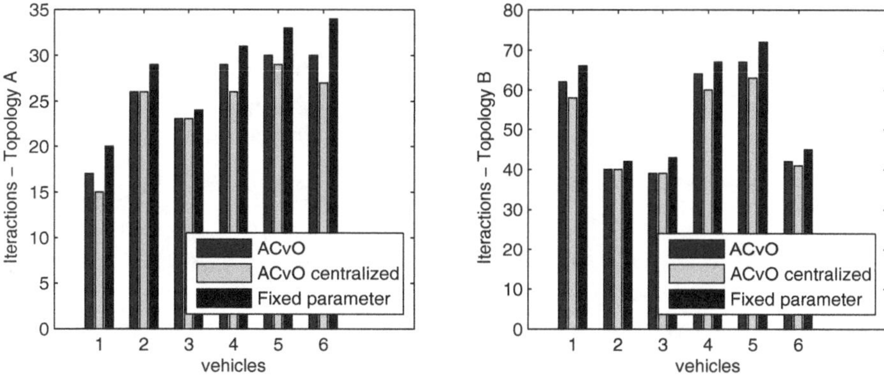

Fig. 5 Tracking consensus convergence iterations for each vehicle of the group

Therefore, we can use our proposed ACvO protocol assuming limitations in the knowledge of the neighbors state without compromising the performance of tracking consensus. (Note that the convergence of the *information states* can be associated to the tracking error).

4.2 On the Computational Time

Since the ACvO protocol is to be used on line, the computational time used to solve the optimization problem should be analyzed. In the simulations performed the sampling time was $\Delta_k = 0.5$ s. Therefore, time consumed to compute the control law should then be less than Δ_k, for all k.

For each iteration k, the algorithm computes the optimization routines sequentially considering synchronized communication. Figure 6 shows the average computational

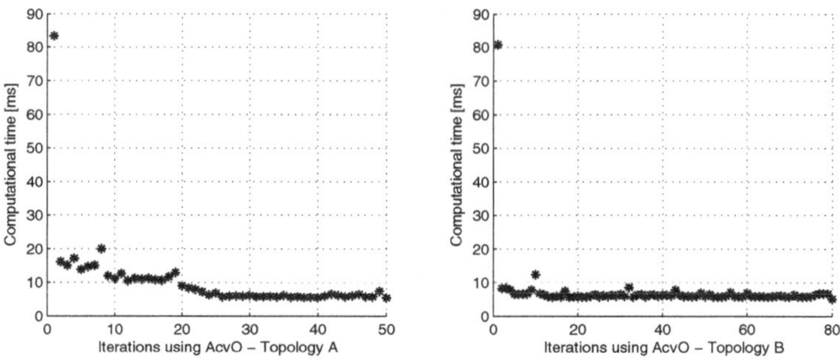

Fig. 6 Average computational time for each iteration of the ACvO

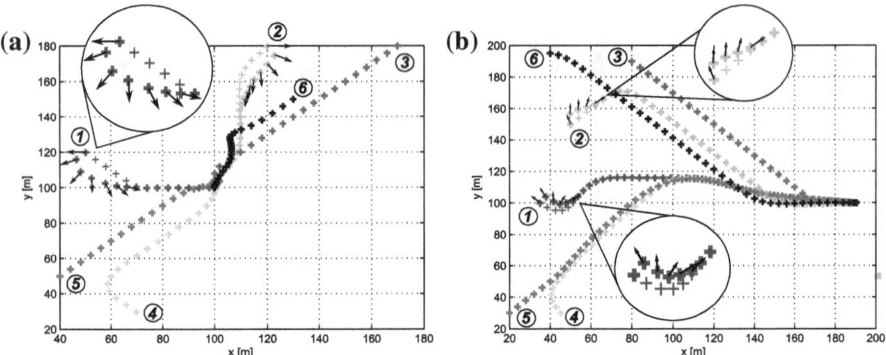

Fig. 7 Dynamics of *information states* ξ_i applying ACvO with integration of mobile robots sensing. **a** Topology A and **b** topology B

time consumed by the ACvO to compute all the 6 control laws shown in the Fig. 4. The sampling time constraint was respected in the previous experiment, using computers with *Intel Centrino Duo T2250* processor clocked at 1.73 GHz and with 2 GB of memory.

It is important to note that the computational time depends on the neighbors cardinality. In Eq. (11) is shown that the control law is directly proportional to the number of jth neighbors vehicles (remember that the control laws are fully decentralized). Another important aspect of the computational time is that the control laws are formulated so that $\Delta_{\xi_i} \geq 0$ and $\Delta_{e_i} \geq 0$ ensuring the convexity of quadratic formulation (12). Therefore, the solution of (12) is comparable to the resolution of a linear programming problem and the optimal global can be found [26].

Optimization of the vehicles sensing angle: Figure 7 shows the results obtained with the application of the ACvO protocol considering the add of the vehicles sensing angle optimization. (the algorithm development was shown in Sect. 3.2).

In the examples, we can see the difference between the behavior of the *information states* when considering the non-holonomic constraint in vehicles 1 and 2. Note that, at the initial iterations, the main objective is to guide vehicle 1, and 2, to the rendezvous point according to the calculated consensus trajectory.

The results presented confirm that the ACvO protocol allows that the *information states* reach tracking consensus. Adding another goal, the vehicle orientation, the result of the group consensus was not compromised and all vehicles were able to perform a trajectory according to their motion constraints.

Optimization with connectivity constraint: Figure 8 shows the results obtained with the application of the ACvO algorithm using the connectivity constraint. This example used topology B and the development of strategy was shown in Sect. 3.3.

This illustrative example shows the influence that the connectivity constraint has on the algorithm implementation. The communication channel between vehicles 1

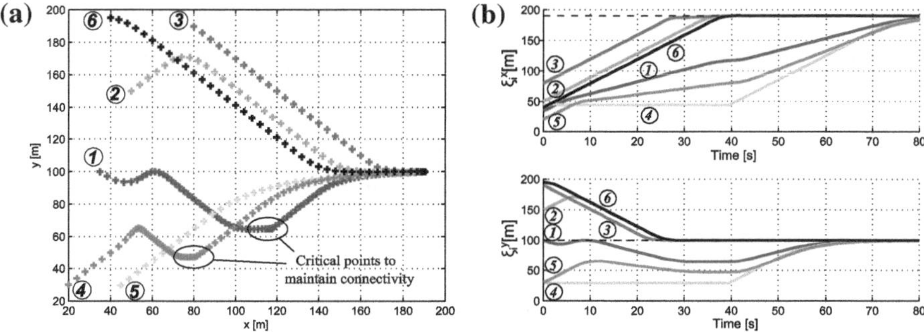

Fig. 8 Dynamics of *information states* ξ_i applying ACvO protocol with integration of connectivity constraint using topology B

and 4 was purposely disabled during the first seconds of simulation. Notice that the vehicles 1 and 5, when reaching the limit distance bound of the connectivity constraint, stop moving and wait for the reestablishment of the a_{14} channel. After recovering communication, the vehicles 1, 4 and 5 continue computing their trajectories to the tracking consensus.

Note that the connectivity constraint has another positive impact. Since the a_{14} communication channel was disabled, vehicle 4 behaves like a false reference to vehicles 1 and 5. However, with the connectivity constraint and communication between vehicles 1 and 2 (link to the other part of the group), vehicles 1 and 5 do not see 4 as a reference.

5 Conclusions

It was presented a methodology for the synthesis of decentralized control laws in order to track trajectories based on consensus. With this approach, using predictive control theory, the cost function was defined by the trade-off between the group cooperation and response requirements. According to this approach, where the addition of terms in the cost function is straightforward allowing more flexibility in design, the methodology presents the following contributions: the first one, the optimization of the sensing range with respect to the rendezvous point, and the second one, the addition of a connectivity constraint to achieve consensus.

The implementation of the ACvO protocol is decentralized and can be used in applications with heterogeneous vehicles in a group with limited knowledge and oriented information flow. The simulations were performed with various scenarios and an evaluation of the results obtained considering the convergence rate performance and computational time confirms that the ACvO protocol allows that the *information states* reach tracking consensus.

In future work we will address time varying topologies and the impact of packet losses inherent to wireless communication in the formulation of the AcvO protocol. Moreover, it is important to establish minimum conditions and correlations between packet losses and the variation of network topology in order to achieve consensus. Another issue that can be explored is the use of the presented formulation based on predictive control applied to other problems, such as formation control, coverage and flocking.

References

1. Olfati-Saber, R., Fax, J.A., Murray, R.M.: Consensus and cooperation in networked multi-agent systems. Proc. IEEE **95**, 215–234 (2007)
2. Schurr, N., Okamoto, S., Maheswaran, R., Scerri, P., Tambe, M.: Evolution of a Teamwork Model, Cognitive Modeling and Multi-Agent Interactions. Cambridge University Press, Cambridge (2005)
3. Murray, R.M.: Recent research in cooperative control of multivehicle systems. J. Dyn. Syst. Meas. Control **129**, 571–584 (2007)
4. Ren, W., Beard, R.W.: Distributed Consensus in Multi-Vehicle Cooperative Control—Theory and Applications. Communications and Control Engineering. Springer, London (2008)
5. Ren, W.: Consensus tracking under directed interaction topologies: algorithms and experiments. IEEE Trans. Control Syst. Technol. **18**, 230–237 (2010)
6. Ordonez, B., Moreno, U.F., Cerqueira, J., Almeida, L.: Generation of trajectories using predictive control for tracking consensus with sensing. Procedia Comput. Sci. **10**, 1094–1099 (2012)
7. Chatterjee, S., Seneta, E.: Towards consensus: some convergence theorems on repeated averaging. J. Appl. Probab. **14**, 89–97 (1977)
8. DeGroot, M.H.: Reaching a consensus. J. Am. Stat. Assoc. **69**, 118–121 (1974)
9. Olfati-Saber, R., Murray, R.M.: Consensus problems in networks of agents with switching topology and time-delays. IEEE Trans. Autom. Control **49**, 1520–1533 (2004)
10. Saber, R.O., Richard, S., Murray, R.M.: Consensus protocols for networks of dynamic agents. In: American Control Conference, Denver, 2003
11. Lin, Z., Broucke, M., Francis, B.: Information flow and cooperative control of vehicle formations. IEEE Trans. Autom. Control **49**, 1465–1476 (2004)
12. Ren, W., Beard, R.W.: Consensus seeking in multiagent systems under dynamically changing interaction topologies. IEEE Trans. Autom. Control **50**, 655–661 (2005)
13. Beard, R.W., McLain, T.W., Nelson, D., Kingston, D., Johanson, D.: Decentralized cooperative aerial surveillance using fixed-wing miniature UAVs. In: Proc. IEEE **94**, 1306–1324 (2006)
14. Xiao, L., Boyd, S.: Fast linear iterations for distributed averaging. In: IEEE Conference on Decision and Control, Atlantis, 2004
15. Decastro, G.A., Paganini, F.: Convex synthesis of controllers for consensus. In: American Control Conference, Boston, 2004
16. Dunbar, W., Murray, R.: Distributed receding horizon control for multi-vehicle formation stabilization. Automatica **42**, 549–558 (2006)
17. Bauso, D., Giarre, L., Pesenti, R.: Mechanism design for optimal consensus problems. In: IEEE Conference on Decision and Control, San Diego, 2006
18. Semsar-Kazerooni, E., Khorasan, K.: Optimal consensus algorithms for cooperative team of agents subject to partial information. Automatica **44**, 2766–2777 (2008)
19. Listmann, K.D., Masalawala, M.V., Adamy, J.: Consensus for formation control of nonholonomic mobile robots. In: IEEE International Conference on Robotics and Automation, New Jersey, 2009

20. Semsar-Kazerooni, E., Khorasani, K.: An LMI approach to optimal consensus seeking in multi-agent systems. In: American Control Conference, St. Louis, 2009
21. Jakovetic, D., Xavier, J., Moura, J.: Weight optimization for consensus algorithms with correlated switching topology. IEEE Trans. Signal Process. **58**, 3788–3801 (2010)
22. Cao, Y., Ren, W.: Optimal linear-consensus algorithms: an LQR perspective. IEEE Trans. Syst. Man Cybern. Part B Cybern. **40**, 819–830 (2010)
23. Kuwata, Y., How, J.P.: Cooperative distributed robust trajectory optimization using receding horizon MILP. IEEE Trans. Control Syst. Technol. **19**, 423–431 (2010)
24. Fax, J.A., Murray, R.M.: Information flow and cooperative control of vehicle formations. IEEE Trans. Autom. Control **49**, 1465–1476 (2004)
25. Ren, W., Beard, R.W., Atkins, E.M.: Information consensus in multivehicle cooperative control. IEEE Control Syst. Mag. **27**, 71–82 (2007)
26. Boyd, S., Vandenberghe, L.: Convex Optimization. Cambridge University Press, Cambridge (2004)

Chapter 3
Localization, Route Planning, and Smartphone Interface for Indoor Navigation

Balajee Kannan, Nisarg Kothari, Chet Gnegy, Hend Gedaway, M. Freddie Dias and M. Bernardine Dias

Abstract Low-cost navigation solutions for indoor environments have a variety of real-world applications ranging from emergency evacuation to mobility aids for people with disabilities. Primary challenges for commercial indoor navigation solutions include robust localization in the absence of GPS, efficient route-planning and re-planning techniques, and effective user interfaces for resource-constrained platforms like smartphones and mobile phones. In this chapter, we present an architecture for indoor navigation using an Android smartphone that integrates three core components of localization, map-representation, and user interface towards a robust and effective solution for guiding a variety of users, from sighted to the visually impaired to their intended destination. Specifically, we developed a navigation solution that combines complementary localization algorithms [10] of dead reckoning (DR) and WiFi signal

This work was done by authors Kannan, Kothari, and Gnegy when they were affiliated with Carnegie Mellon University.

B. Kannan (✉)
GE Global Research, Niskayuna, NY 12309, USA
e-mail: balajee.kannan@ge.com

N. Kothari
Google Inc, Mountain View, CA, USA
e-mail: leeglechan@gmail.com

C. Gnegy
University of Pittsburgh, Pittsburgh, PA 15213, USA
e-mail: chet.gnegy@gmail.com

H. Gedaway · M. F. Dias · M. B. Dias
Carnegie Mellon University,Pittsburgh, PA 15213, USA
e-mail: hend.gedaway@gmail.com

F. Dias
e-mail: mfdias@ri.com

B. Dias
e-mail: mbdias@ri.com

A. Koubâa and A. Khelil (eds.), *Cooperative Robots and Sensor Networks*, Studies in Computational Intelligence 507, DOI: 10.1007/978-3-642-39301-3_3,

strength fingerprinting (SSI) with enhanced route-planning algorithms to account for the sensory and mobility constraints of the user to efficiently plan safe routes and communicate the route information with sufficient resolution to address the needs of the users. To evaluate the feasibility of our solution, we develop a prototype application on a commercial smartphone and tested it in multiple indoor environments. The results show that the system was able to accurately estimate user location to within 5 m and subsequently provide effective navigation guidance to the user.

Keywords SSI-based localization · Path-planning · Assistive technology · User interface

1 Introduction

The ability to independently navigate urban environments is a fundamental necessity for all of us. While outdoor navigation is a well established field, analogous techniques for indoor environments are still in their infancy. The problem of indoor navigation is further complicated when applied to a wider user community than that of just sighted users. In general, user type can be classified into two main categories; those that are sighted and those that have visually impairments, including partial blindness, deafblind, etc. Solutions for users with low or no vision require a higher accuracy of pose estimation, path planning with constraints and non-visual, context sensitive user interfaces. There is a need for a low-cost, practical, customizable, and easily deployable mobility aid to safely navigate indoor environments for a variety of users, from the sighted to the visually impaired and even the deafblind.

In general, existing indoor navigation tools [1, 7] have a few critical shortcomings. First, there do not exist low-cost localization solutions that can provide a high resolution of accuracy required for navigating the different user groups. Second, most of the navigation technologies do not provide routing instructions based on landmarks about the environment. Further, flexibility and customization of elements based on user preference is not explored in depth, a critical need for navigating visually impaired users. In summary, there is a lack of low-cost, accessible, customizable, and user-friendly navigational aids to allow sighted and visually impaired users to safely navigate or evacuate buildings.

In this chapter, we outline an innovative smartphone-based prototype solution for indoor navigation that combines intelligent robotic systems, WiFi-based signal strength estimation, advanced path-planning, and multi-modal interface towards improving the overall safety and reliability of navigation for a variety of end users. The uniqueness of the solution lies in its capability to handle static and dynamic information, application of robotics technology for data collection, WiFi-based signal strength techniques, multi-level path-planning, and multi-modal user interface for navigating indoor environments at relatively low costs. To our knowledge, this is the first architecture that attempts to integrate the core navigation components of

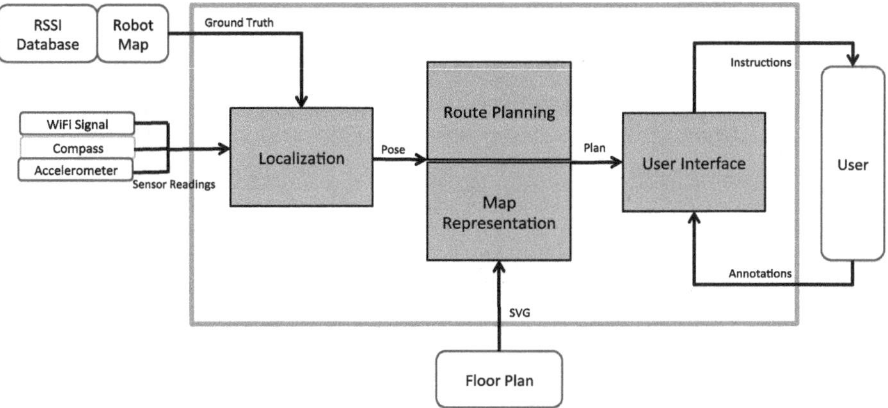

Fig. 1 Indoor navigation system architecture

path-planning and localization with a multi-modal interface towards a navigation solution for sighted and visually impaired users.

The rest of the chapter is organized as follows. In Sect. 2 we introduce our indoor navigation solution and identify the three key research challenges that need to be overcome. Sections 3, 4, 5 detail our approach to overcoming the localization, path-planning, and user interface challenges respectively. In Sect. 6 we detail the experiments performed to validate the developed system and analyze the obtained results. Finally, in Sect. 7, we conclude the chapter with a summary of the results and outline future work directions.

2 A Smartphone-Based Indoor Navigation Solution

Current mobile phone technology has evolved to a point where affordable smartphones with a variety of sensors are readily available to the public. Ever improving features of affordability, ubiquity, and portability, increased sensory and computational power along with low power consumption fueled by readily available batteries, have opened up a number of interesting applications.

Our indoor navigation architecture consists of three major components: indoor localization; path-planning and map management; and user interface, as illustrated in Fig. 1. To achieve effective localization in the absence of GPS, we combine complementary localization algorithms of dead reckoning and WiFi signal strength fingerprinting. These measurements along with pre-built maps of the environment are combined using a particle filter for robust pose estimation. Towards effective path planning , we use a hierarchical map representation combined with an iterative path-planning algorithm for providing fast and efficient user specific routes. The focus of the multi-modal interface is on better quality instructions and enhancing user flexi-

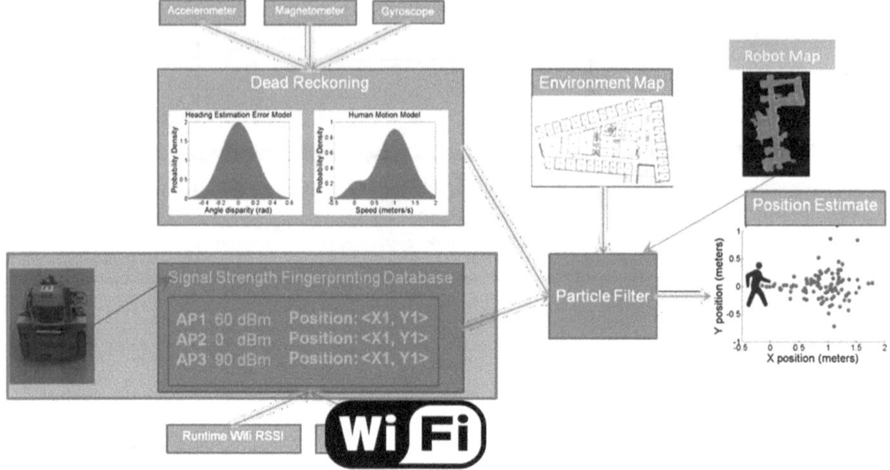

Fig. 2 Localization solution

bility in terms of the verbosity of information presented and the different input/output modalities, including visual, audio, gestures, and haptic feedback.

These modules run independently and continuously exchange information as new data comes into the system. The data flow in the system originates with the localization module receiving information from the various sensors. The raw data, from WiFi and dead reckoning, is combined using a particle filter to provide an estimate of the user's location indicating the dispersion of probability of the user's location within a building. The path-planning component, in turn, uses the estimated location for its path-planning algorithm and generates an updated route for the user. The User Interface (UI) component translates the generated route into user specific instruction set represented via a visual or auditory representation.

3 Indoor Localization Solution

Some of the more commonly used localization systems include GPS [13, 16], Wi-Fi or Bluetooth devices [19], ultrasound [16], and Passive Radio Frequency Identification (RFID) tags [14, 15]. Technologies such as GPS require clear line of sight and are not suitable for indoor navigation. Others such as RFIDs require pre-installed infrastructure and extensive changes to the environment, are inherently expensive, and do not provide the requisite resolution of accuracy.

While the short range properties of Bluetooth and Near Field Communication (NFC) have been used to constrain the estimate of the users location [5] they have the drawback of requiring installation of markers in each environment where the solution is used. SSI fingerprinting measures [12] for pose estimation is an accepted and popular technique for indoor pose estimation with an accuracy range of 3–10 m. A drawback of existing SSI-based solutions is that they are geared towards devices with significant computation capabilities and high-fidelity sensors. Further, current methods for building a SSI calibration database is tedious, labor-intensive, and requires a large number of samples [2]. Alternately, a DR system comprising of accelerometer, magnetometer, and gyroscope sensors can provide fast and accurate estimation of local pose [8]. While effective over short distances, DR solutions have the drawback of being local estimation techniques and have to be seeded with an accurate initial position for valid estimation. Furthermore, over time and distance the sensory error accumulation is unbounded. particle filters (PFs) are commonly used techniques for integrating disparate information sources from multiple sensors towards robust localization [4]. PF methods can readily incorporate nonlinearities introduced by WiFi signal strength filtering as well as handling restrictions imposed by the robot map and can track multiple hypotheses by dividing particles proportional to their likelihood.

For our solution, we select the WiFi radio combined with dead reckoning using accelerometer, magnetometer, and gyroscope sensors for pose estimation. We build on the general ideas of DR and SSI fingerprinting, from which we derive a baseline implementation using a PF and a k-nearest neighbor approach. A gait-based motion model combined with a heading estimator provides a pre-filtered dead-reckoning sensor estimate to the particle filter (PF) (see Fig. 2). Simultaneously, pose is estimated based on fingerprinting between observed WiFi signal strength readings and robot-based pre-collected database of SSI estimates. The combined sensor data is fused and filtered using a PF which results in a smooth and continuous pose estimation state. While the notion of combining WiFi fingerprinting with dead-reckoning using either a particle filter or Kalman filter has been explored priorly [18], the uniqueness of our approach lies in the implementation of the complementary nature of the solution as well as in the efficient adaptation to the smartphone platform at a relatively low computation cost.

3.1 Dead Reckoning

The orientation is tracked using two complementary methods. The first is to employ an accelerometer and magnetometer to give a reference direction for gravity. The benefit of this method is that each measurement is externally referenced, so orientation errors do not accumulate over time. The drawback is that magnetic anomalies which interfere with the operation of the compass are common in indoor environments. The second method is to use a gyroscopic sensor that is much less noisy and is not susceptible to external interference. However, as they measure angular velocity rather than angular position, only the relative movement of the phone can be derived from the gyroscope readings, and the error in the orientation tracking accumulates over time without a bound. To robustly determine heading, these two methods are merged.

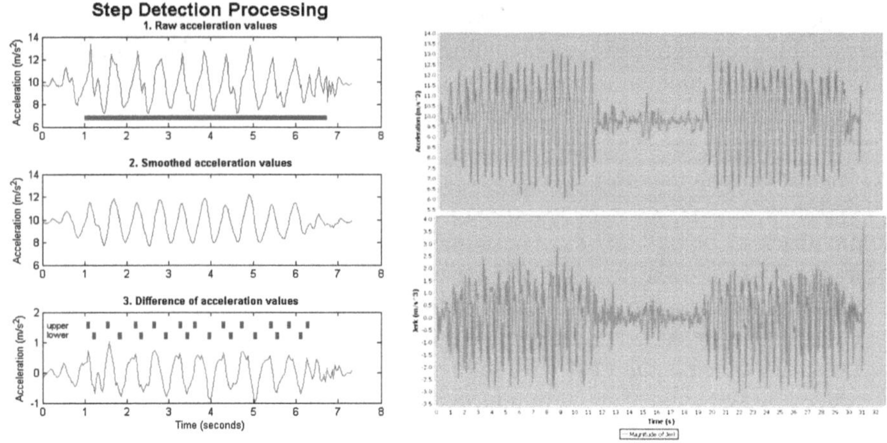

Fig. 3 Accelerations during a short walk

The inputs to the combined filter are raw accelerometer, magnetometer, and gyroscope readings as they are received. The output is an estimate of the azimuth, pitch, and roll of the phone in a global frame. If the magnetic field strength, observed from magnetometer readings, deviates significantly from the strength expected for the users approximate global location, the data is filtered out. Further, the gyroscope data is integrated using the trapezoidal method to capture the change in angular position in the local coordinate system. Once calculated, the change in angular position is converted into the global orientation frame. If an estimate of the transformation from the gyroscopes local frame to the global orientation frame is already available, this can be done by applying the transformation matrix R to $\delta theta_{local}$

$$\delta\theta_{local} = (1/2gyro_{clean}(t) + gyro_{clean}(t-1)) * \delta t \tag{1}$$

$$\delta\theta_{global} = R\delta\theta_{local} \tag{2}$$

$$\theta_{global} \leftarrow \theta_{global} + \delta\theta global \tag{3}$$

Occasionally, θ_{global} must be re-initialized from the accelerometer and magnetometer data to limit the accumulation of error from the gyroscope integration.

We use two different approaches to detect movement, both of which exploit the periodic nature of the walk cycle. The first uses a peak detection filter with alternating high and low thresholds to detect individual steps. In order to detect steps accurately, we smooth the accelerations using a running average filter and then remove offset errors and drifts by taking the first difference of the smoothed result. The second method, which has the advantage of detecting movement continuously in time rather than at the step level, looks at the variability of the acceleration readings, with the assumption that higher variability corresponds to movement. The standard deviation of the acceleration values over a fixed time span is used to detect movement. The size of the time span and the threshold that must be met to report movement both contribute to a tradeoff between the delays that are incurred before the filter is able to detect movement, and the potential for false positives.

Despite the fast estimation time and potential, the DR system is limited in performance. DR solution is a local estimation technique and has to be seeded with an initial position for valid estimation. The on-board gyroscopic sensor, while effective over short distances, accumulates error and drifts un-bounded over time. Further, the magnetometer can be easily distorted by magnetic anomalies, which are relatively common in cluttered indoor environments.

3.2 Signal Strength Fingerprinting

To compensate for the drawbacks of the DR system, we use a WiFi-based signal strength fingerprinting approach. The signal strengths to several WiFi access points (APs) measured by the phone at run time are compared with a signal strength map generated earlier. The difference between the expected reading for that position in the map and the actual signal strength reading is used to adjust the weight of the particle. We use a Euclidean distance metric in signal space between the APs common to both readings. Effective distance is calculated as a weighted average of the nearby calibration points to reduce noise. In addition, a penalty is imposed for APs that were disjoint between the two readings in proportion to the signal strengths of those APs. Additionally at runtime, WiFi signal strength fingerprinting is used to initialize the system and provide a rough global location estimate.

To perform the SSI fingerprinting, it is necessary to create a database of signal strength information from the environment correlated to a free space map of the environment. This database forms the basis for comparison and subsequent estimation. Collecting this information by hand is laborious and prone to error, so we developed an automated solution that uses a pioneer robot equipped with a SICK LMS200 laser rangefinder, a fiber optic gyroscope, and a mounted smartphone, to collect the signal measurements. The phone collects signal strength information continuously over the course of the automated run. At discrete intervals ($\sim$1 m), the signal strength readings from the phone are correlated with the current position of the robot. The robots on-board sensing allows for a continuous and accurate estimation of its pose. At the same time, the robot also builds a $2-D$ map of the environment using the laser rangefinder. The result is an accurate, high density sample of signal strength information in a short amount of time. Further, the shape and structure of the laser map allows us to speed up our pose estimation and reduce computation by discarding PF particles that lie outside the bounds of the map. Figure 4 show the robot setup and subsequent pose estimation.

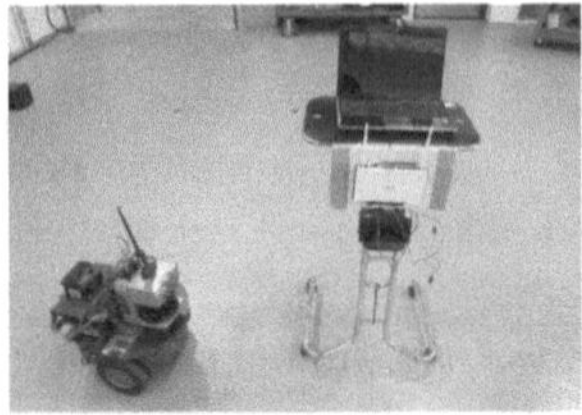

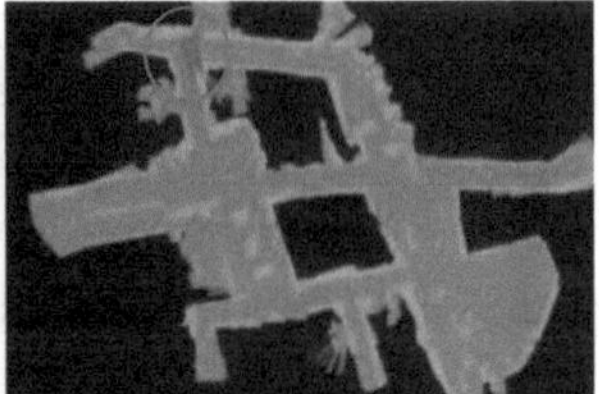

Fig. 4 SSI database generation using pioneer robot and subsequent PF-based pose estimation

3.3 Particle Filter

The use of a PF is critical for minimizing the accumulation of localization error over time. PF's use sampling methods to track many possible hypotheses, updating them every time new information becomes available, and can be adapted to handle non-linearities and non-gaussian noise models. Crucially, the performance of the filter can be scaled with available computation power by varying the number of particles that are tracked.

A PF has three major components: a motion model that updates the positions of particles, an observation model that sets particle weights, and a re-sampling algorithm for modifying distribution to reduce variance. For our system, the motion model is derived from the DR algorithm, with the heading and movement speed extended into a spatially directed Gaussian distribution. The external measurements used by the filter to update particle likelihoods come from two different sources. The first is the robot map of the environment. Whenever a particle enters a space that is indicated to be impassable in the map, its likelihood is reduced to zero. The second source of external measurements is the aforementioned WiFi signal strength measurement. We use the importance sampling algorithm to resample particles. This method reduces the variance of the particle distribution by eliminating unlikely particles and duplicating particles that are more likely. Particles that lie outside the constraints of the environment are weighed lower than others. Specifically, the new particle distribution is formed from the old one by selecting particles (with replacement) with probability proportional to their weight. This focuses the particle distribution on the area of maximal interest (see Fig. 4).

4 Map Representation and Path-Planning

Hierarchical maps are a popular technique to represent environments in dynamically changing domains such as robotics [3], reducing the search space to the sufficient sub-graph(s) needed for search. In our approach, we use a hierarchal representation to encapsulate the environment at varying granularities (see Fig. 6). High-level maps are used to represent larger areas of a building without representing the exact spatial

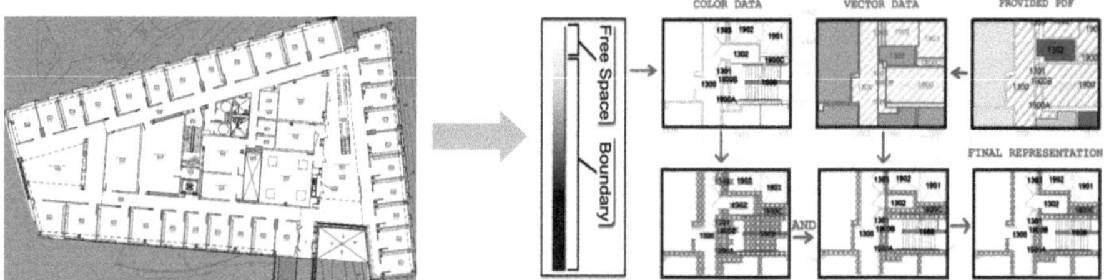

Fig. 5 SVG-based automated map translation

relationship of individual locations inside rooms and corridors, while low-level maps are used to represent individual rooms with enhanced spatial detail. Consequently the high-level maps are used to construct abstract plans to navigate between floors and rooms of a building, whereas the higher resolution of the low-level maps are essential for precise navigation within rooms and hallways. This multi-level approach allows for multi-floor planning and can be extended to planning between buildings. Topologically, in our system the high-level maps are represented as a graph and the low-level maps as a grid structure connected to the leaf nodes of the graph.

4.1 Automated Map Translation

In order to model the indoor environment and extract a logical layout of the building, we utilize a vector graphic representation, the Scalable Vector Graphics (SVG) format, of the floor plans of the building. SVG is an XML-based format that can be processed by a multitude of standard tools in order to extract data from an image. Using the defined formats, an interconnected network of rooms is established based on their adjacency information and a graph of rooms and connecting hallways is built (see Fig. 5).

4.2 Path Planning

An effective route planner should have both high throughput and low delays and should be capable of fast dynamic re-planning. D*lite algorithm [9] is one such algorithm capable of planning paths in changing environments and enabling rapid re-planning if changes in travel costs are discovered. Cagigas [3] introduce a new hierarchical extension of the D* algorithm for robot path-planning, where a down-top strategy and a set of pre-calculated paths are used to improve performance. The outlined approach has the drawbacks of needing pre-computed paths and operating on a uniform granularity of information representation across the different nodes of

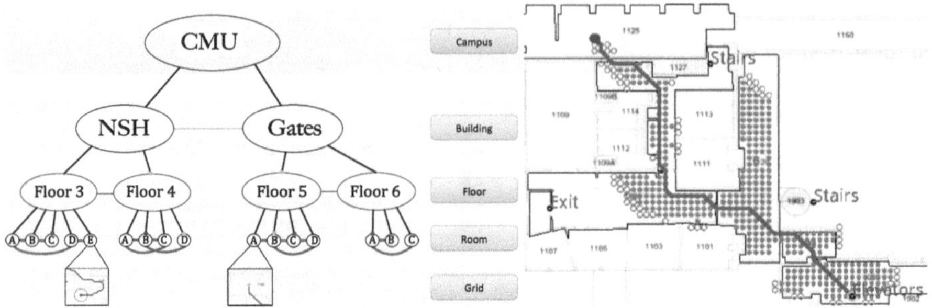

Fig. 6 Hierarchal path-planning

the graph. We extend the hierarchical D* approach and outline a top-down multi-level planner for varying map granularities.

4.2.1 High-Level Planner

The high-level A*-based planner handles building-wide connections accounting for exits, transitions points between floors (stairs, elevators), and planning across multiple floors. Further, the planner provides a restricted path for the grid planner, constraining the search space and consequently reducing the computational overhead. Based on user-location and the intended destination, the planner quickly searches the graph hierarchy, above the user and destination levels, until it finds a connecting node. After a route is established, the planner recursively searches for the shortest path through the network of edges and vertices moving down one step at a time to verify connections.

4.2.2 Grid Planner

Once the high-level planner is run and restricted path generated, we then run an instance of a D^*lite grid planner. By constraining the search space of the grid planner to only the generated path, we reduce the computational overhead of the grid planner. D^*lite is a dynamic path-planner capable of handling changing environments in an efficient, optimal, and complete manner. The planner operates at the individual floor level and is used to find a fine-grained path to either the destination (should it be on the same floor) or an exit (if the destination is on another floor). This improves efficiency as the destination node is fixed, but the start node varies with the change in user's location. Specific to our system, at the lowest level of the hierarchy there are rooms which correspond to vertices used in the D*lite algorithm. These rooms are connected by doorways, which correspond to the edges used in D*lite (see Fig. 6 for illustration). The hierarchy extends upwards to the floor level and building level, with floors and buildings corresponding to the vertices while exits (staircases and elevators) and roads corresponding to the edges at each respective level.

The planner has two core lists, *OPEN* and *CLOSED*, that determine whether the node has been added to the tree and subsequently evaluated. Optimality is determined by identifying the node, M, in *OPEN* that minimizes a defined cost function, $f(M)$, while ensuring all neighboring nodes to M are not blocked by a barrier. In the event that two nodes have an equal cost, a shortest-distance-to-goal heuristic is used for node selection.

We further optimize the grid planner for the smartphone platform by modifying the re-planning step of the algorithm to actively time-stamping the nodes that are moved into the OPEN or CLOSED lists. When re-planning after time T, we can quickly roll back nodes with a *CLOSED* time-stamp greater than a T, re-assign them to the *OPEN* set, remove all node with an *OPEN* time-stamp greater than T, and re-run the planners. This significantly reduces the computational overhead for re-planning. The adaptations to the search space and re-planning allows us to significantly speed-up the process of path-planning, especially across multiple floors.

5 User Interface

The quality of user interaction with a navigation system often dictates its usefulness. Designing user interfaces for sighted users is a well understood problem, with solutions ranging from Google Maps to stand alone navigation units. However, those who could perhaps most benefit from additional assistance during independent navigation, the visually impaired, are often unable to access many of these tools. Most navigation user interfaces are graphically oriented and are not accessible to people with visual impairments. Buttons, vibration patterns, metaphors, spatial sounds, spatial language, virtual displays, gestures and speech recognition are common input modalities for the visually impaired [6, 11, 14, 17]. The main limitations of the existing navigation technologies are the quality of the instructions provided and the usage flexibility. The interaction modality is often pre-determined and the user does not have the flexibility to easily switch between different modalities depending on the situation. Giving context about the surrounding environment and allowing users to customize the extent of output information detail is yet another essential missing feature.

5.1 Needs Assessment

A detailed needs assessment process is a key component for developing a functional user interface for the visually impaired. The process of needs assessment involved gathering information about suitable interface design, instructions and landmarks that are important for navigating a visually impaired user in an unknown indoor environment. The needs assessment process was conducted in two phases with a sample pool of 20 survey respondents, 18 of whom are blind or visually impaired, and two of whom are individuals without visual or hearing impairments, over several months. The process involved initial interviews with visually impaired individuals and experts in the Orientation and Mobility (O&M) field and detailed surveys and interviews with visually impaired and others at local community organizations. The purpose of the needs assessment were twofold:

- Encapsulate the needs and preferences of users based on literature reviews, past projects, and experiences.
- Identify additional information targeted towards blind, deafblind, and other people with disabilities.

The observations from phase 1 helped us gain a better understanding of the landmarks of interest to potential users, while initial interviews provided feedback on input/output preferences for interacting with technology devices. Interviews revealed the difficulty of emergency evacuation in large unfamiliar indoor buildings involving complex turns and intersections. Majority of the users prefer audio as the preferred output modality, while gestures are the preferred input modality. Further, customizability of the interface is a key functionality that needs to be incorporated when designing the interface. Finally, providing context sensitive navigation instructions, such as environmental cues and landmarks, is essential for effective navigation.

5.2 Multi-Modal User Interface

Drawing upon notions of universal design, we have developed an accessible tool for a variety of users ranging from sighted to visually impaired and even deafblind individuals. The key characteristics of the interface are an improved instructions set and enhanced user flexibility in terms of the verbosity of presented information and the different input/output modalities. The interface maintains a good balance between the quality of the navigation instructions and the automatic generation of these instructions. To ensure the planned path is converted to context sensitive verbal instructions, we have implemented an automated translator. The interface supports three levels of routing instruction details: high (H), intermediate (I), and low (L). At the high level the user is guided at the hallways and intersections level until final hallway, at which point a count of doors along the way to the destination is given. The intermediate level adds step-by-step directions and a count of doors along the whole route. The low level adds additional landmarks and contextual information about the environment.

The interface supports different input/output modalities, including visual, audio, gestures, and haptic feedback. To ensure reliable usage, even in crowded noisy environments, we use gestures as a tool for navigating the system menu options. The user can use one of five defined gestures (see Fig. 7) on the screen to navigate the tool. The gestures list allow the user to obtain the next instruction in the list, pre-

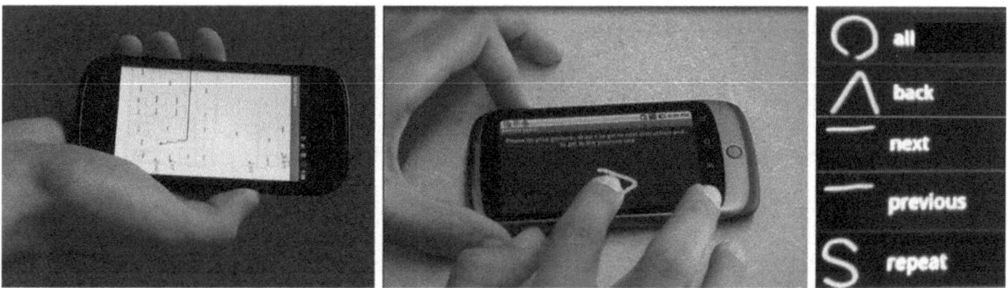

Fig. 7 Multi-modal interface (visual interface on the left and gestures-based interface on the right)

Fig. 8 Sample indoor user test environment

vious instruction, repeats the current instruction, gives the user the complete route instructions, and allows the user to navigate back to the main navigation menu.

To ensure flexibility in the use of contextual information, the user is provided with options to add and delete annotations. The annotations could be obstacles encountered as well as landmarks recognized during navigation. Further, the use of landmark based contextual information allow for ease in translation of information when requesting help from a sighted user. When the user selects a landmark, the current location of the user is retrieved from the localization component and the map tagged with the selected landmark.

We augment the audio output with haptic feedback for improved user interaction, especially in crowded environments. We define a set of vibration patterns differing in the number and the length of vibrations. The vibration patterns are associated with keywords that are common to different instructions. The keywords are grouped into four different groups, with each group having a different number of vibrations in its pattern. The first group contains keywords that relate to directions, the second group relate to motion commands, and the third group provides information about the environment, and the fourth group captures special cases.

6 Experiments and Results

We implemented the developed solution as a stand alone application for the Google Android platform and tested the individual components as well as the integrated system in multiple indoor environments. The primary issue encountered during development was load balancing between the different system components. Consequently, the resource intensive components of the system were developed as background services integrated using the main application that displays the user interface. The user interface provides the user with options to toggle path-prediction, select a destination, add annotations, manually re-plan the route, etc. Data communication between the different modules is handled via Android's broadcast system. Using this system, a service sends a broadcast update that can be accessed by any registered application. This push-based data service affords an added layer of robustness to the system, in the event of computation back-log. In such an implementation, each of the different modules acts as an independent application depending only on periodic communication to further its own state.

We tested the integrated system with visual interface on a single floor of an indoor environment with a set of three sighted users. Three sets of experiments involving navigation from a defined start to multiple destination locations were performed. Each set of experiments was repeated multiple times for consistency, for a total of 20 runs. There were approximately 20 WiFi access points observed in the environments. In all experiments the user starts from a pre-defined starting location and has two potential destination locations to head towards. The participants were instructed to hold the cell phone device in their hands pointing forward and to walk at their normal pace. The routes and destinations are illustrated in Fig. 8 and we tabulate the results from our experiments in Table 1.

6.1 Indoor Localization Tests

The system is not initialized with a starting location and attempts to converge based on the measured signal strength. From the table we can see that for the localization algorithm, on average, is able to converge to within 1 m of the actual starting location for all three experiment types. Paths one and three are approximately 15 m long, while in experiment two the user travels a total of 32 m to destination. Moreover, we can see that the mean error for the localization over the course of the different experiments is approximately 3 m and in most cases the system is also able to converge to a final destination position within an error bound of approximately 5 m. It is interesting to observe that the estimate of final position is of a value greater than the calculated mean error. We believe this is due to the fact that the selected destination locations lie outside the boundaries of generated robot map. Consequently, there exists a smaller sample of data points correlating to the destination locations in our SSI database, leading to a higher uncertainty in pose estimation.

Table 1 Experimental results

Environment	Starting location	Exp.	Path-length (m)	Start pose ϵ (m)	Final pose ϵ (m)	Mean ϵ of (m) most prob. loc. (m)	Time to predict (s)
I	(0,0)	1	16.5	0.71	3.42	2.20	43
I	(0,0)	2	31.67	0.94	4.26	2.49	53.47
I	(0,0)	3	15.46	0.90	3.34	2.71	30.324

To better understand the quality of localization, we performed additional closed loop experiments in which the participants walked a path through hallways delineated with cones. Four individuals participated in the test for environment 1 and three for environment 2. Each set of experiment was repeated multiple times for consistency. The total length of the traversed path in environment 1 was approximately 120 m, and the path in environment 2 was approximately 72 m, with 30 WiFi access points in both environments. The DR information was obtained at a faster rate (30 Hz) than the WiFi signal measurement (1 HZ). Based on our observations, the signal strength variations within 1 m are small enough that they can be effectively ignored. Additionally, the PF-based posed pose estimation is inversely proportional to the distance between the measured points, i.e., as the distance between points increases, the uncertainty in estimation becomes larger leading to a coarser positioning solution. Consequently, we settled on a distance of 1 m between our collected SSI data points. As part of our future work, we will attempt to better define the signal strength propagation over short distances and its affect on PF performance.

Further, we compared it against an offline standard PF solution as well as against the performance results outlined by Wang et al. in [18]. In our offline system, sensor data collected from the above runs were passed through the PF and post-processed offline on a laptop. In contrast, the online version of the filter was run on the device and the resulting output tabulated. The limited processing capabilities of the Android handset required major modification to the PF algorithm used in the off-line version. Specifically, the updater (DR) portion of the PF has been separated so that the orientation accuracy can still be maintained while the PF processes the WiFi data. Further, the online version runs on a much more sparse data so as to not overwhelm the system, whereas the offline implementation has the advantage of running on the full set. Additionally, data from runs (offline and online) suffering from significant magnetic distortion was disregarded. Figure 9 illustrates the mean error is position over varying path distances collected over two indoor environments.

Over the relatively short duration of the experiments, the DR method was able to track the participants with relatively low error and outperformed the WiFi only estimation. The offline DR had a mean path error of approximately 5 m, whereas the WiFi approach error was on the order of 10–15 m. This is somewhat misleading, because of two factors. First, in both instances, the PF was manually initialized to the correct position for the DR-only results, whereas it was given a uniformly random initialization in the trials that used WiFi. Having a confident estimation of the starting location is not a given for all situations. Ideally it would not be necessary

	Mean Error of Most Probable Location		**Additional Features**	
	Environment 1	Environment 2	Auto Init. Pos.	Bounded Error
DR Only	5±3 meters	6±2 meters	No	No
Wifi Only	15±10 meters	10±4 meters	Yes(<5 seconds)	Yes
DR and Wifi (offline)	5±3 meters	7±2 meters	Yes(<5 seconds)	Yes
DR and Wifi (online)	3±3 meters	5±4meters	Yes(<5 seconds)	Yes

Fig. 9 Comparative results

to manually initialize the filter, as this places an additional burden on the user of the system. Second, DR is known to drift unbounded over time. Finally, a DR-based method that uses magnetometer is susceptible to magnetic anomalies prevalent in the environment. This is emphasized in Fig. 9, where the increase in mean error over path length can be attributed to drift in the dead-reckoning solution.

Further analysis of the collected data revealed that the offline PF takes about 0.5 s per processed reading whereas the online system can only output a reading about once every 1.5 s. Interestingly, the online system shows at least as good a performance as the offline system, the laptop-based PF method implemented by Wang has a mean error of 4.3 m, while our offline PF method has a mean error of 5 m. Additionally, both the online and offline systems had similar closed loop error measurements ($\sim$6 m) averaged over the different runs. Our reasoning is that while the higher computation of a laptop allows the offline system to process the data at a much faster rate, it introduces an increased sensitivity to sensor variation, especially from WiFi readings, leading to a reduced performance. On the other hand, the lower processing rate of the online system results in the error growing at a much slower rate leading to an improved estimation.

We can see that our localization solution performs as well as the alternate solutions developed for higher computational platforms with higher-resolution and dedicated sensors. The online system further highlights the robustness of the developed system if one of the components is unavailable, by simply not feeding that measurement to the PF. On the other hand, it is important to note that the tight constraints based on robot map and the reduced the number of particles in operation, make the filter sensitive to change. Consequently, once error accumulates in the system it takes longer to dissipate than for the traditional PF-based systems.

6.2 Map Representation and Path Planning Tests

The SVG maps used in the experiments were about 1100 $\times$ 800 pixels with a floor space of 250 $\times$ 140 m^2. For the grid planner, we use a defined grid cells of about 1.4 m^2. We further scaled down the map by a factor of 6, which resulted in 6 $\times$ 6 cell grid-to-pixel ratio. Towards boundary determination, each 6 $\times$ 6 cell of the map is checked for 8-bit grayscale intensities less than a pre-defined threshold.

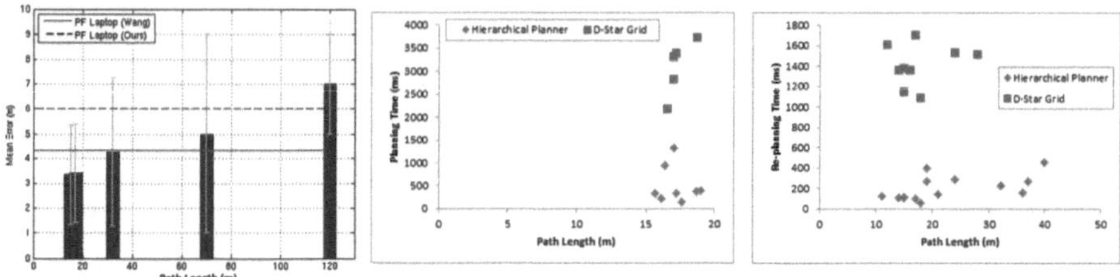

Fig. 10 Localization and path-planning test results. From *left* to *right*: comparing localization performance for implemented solution against the one outlined in Wang et al., comparing initial planning time for different path lengths, and comparing re-planning times for different path lengths

Table 2 Map loading and path-planning times

Env.	With localization and prediction	Process time (s)	Load time (s)	Grid planning (ms)	Grid re-planning (ms)
I	Yes	46.90	3.58	749	22.8
I	No	46.90	3.44	408	23

Table 2 outlines the profiled times for loading the environment map and for running path-planner on it. For path-planning, we compute a path stretched across the entire map. From the table we can see that the time taken to process a new map of relatively high-granularity is around 47 s. Looking at the path-planning component we can see that while it takes a relatively large amount of time for the initial plan, future re-planning is significantly faster.

Figure 10 compares the performance of the developed hierarchical planner against a regular D^*lite planner. Specifically, we look at initial planning time as well as re-planning time for a variety of path-lengths. From the figures we can see that the hierarchical planner performs significantly better than a standard D^*lite planner. The time taken to process a new map is around 500 ms, while re-planning time averages around 200 ms. In the case of the D^*lite planner, initial planning times are on the order of 3,000 ms and subsequent re-planning time averages 1,400 ms. After the initial processing, future load times and subsequent re-planning drop dramatically, as once a connected graph is built, we make the assumption that the base topography (as indicated by the floor plan) remains the same (see Table 2). Consequently, we load only the connectivity graph reducing overall load time. Despite the added complexity, our re-planning algorithm is able to significantly reduce re-planning time, highlighting the effectiveness of our map-representation and path-planning.

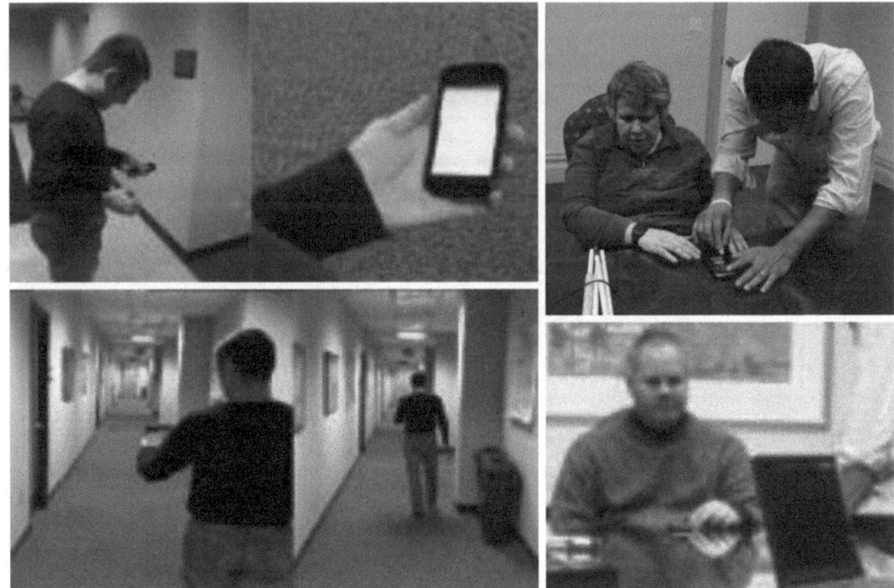

Fig. 11 User navigation tests in multiple environments (sighted user tests *on the left* and visually impaired user tests *on the right*)

6.3 User Interface Testing

In stage I of user testing, we evaluated the integrated system with sighted users. By performing initial tests in a controlled environment, we can evaluate the developed application without endangering the visually impaired users in the event of system failure. The interface provides the user a visual representation of the collated information regarding current position, destination, and the generated route overlaid in a map of the floor (see Fig. 11). In all the experiments, the users were able to reach their destination following the on-screen instructions.

In stage II of the user testing, a total of eight visually impaired subjects were asked to follow a trial route to a set of three pre-defined destinations using the instruction set generated by the interface. The primary goals of the testing process were to evaluate the interface usability and the quality of the instructions provided. For the test, the users had the flexibility to customize different levels of instruction details for the three routes. The observer took notes on what user and system performance, quality of solution, and ease of interaction.

From Fig. 12 we can see that while the navigation tool had a relative high success rate, some participants had difficulty getting to goal destination. While the observed success rates for reaching the goal were pretty high (83 %), there were a few caveats. For routes 2 and 3, one of the participants got to goal with some assistance of the tester, while in two other cases the participants required a do-over. There were four

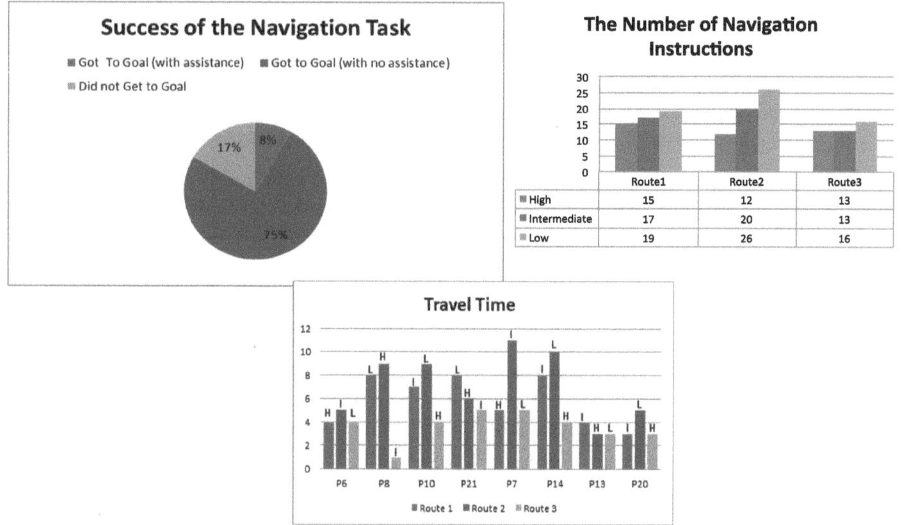

Fig. 12 User interface results in terms of successful navigation, number of instructions needed for navigation, and time take to complete navigation tasks

cases where participants failed to get to goal, with three of the cases on route 1 and one on route 2. For the three cases in route 1, one was with high level instructions, one was with intermediate level, and one was with low level, and the one case in route 2 was using low level instructions. Interestingly, two of the cases were with a single user who had issues switching over the instructions. The number of instructions that the user has to follow seems to have the main influence on the travel time.

The overall feedback for the application and the interface was overwhelmingly positive. Most users liked the use of tactile feedback for noisy environments to overcome the limitations of audio, as well as the layout of the menu options and were able to relate to the logic behind them. There was also positive feedback on the flexibility the menus provided, especially as it pertained to annotations, and the use of gestures as an input option.

7 Conclusion and Future Work

In this chapter, we have outlined a smartphone based indoor navigation solution that addresses challenges pertaining to indoor localization, path planning, and effective user interface for sighted and visually impaired users. The indoor localization system uses dead-reckoning and WiFi fingerprinting for providing way-finding directions. We combine the localization solution with enhanced route-planning algorithms to account for the sensory and mobility constraints of the user to efficiently plan safe routes and communicate the route information with sufficient resolution to address the needs of the users. To evaluate the feasibility of the developed solution, we implement a prototype application on a commercial smartphone and tested it in multiple indoor environments. The results show that the system is capable of providing effective navigation guidance to the users.

There are several interesting directions to explore in future work. The limitations of WiFi signal estimation can be improved by incorporating alternate sensors into the localization framework. Effectively representing large scale floor plans and developing n-level hierarchical map-representation and corresponding path-planner is an interesting and open problem. User interface issues such as simpler gestures, clearer audio feedback, enhanced landmark list, having a horizon feature, and customizable level of instructions need to be addressed towards a truly accessible interface.

Acknowledgments This work is sponsored in part by the Google Core AI gift from Google Inc., The Boeing Company, Google Professor Partnership Award, CMU's Traffic 21 Grant, Berkman Award, and CMU-Qatar Faculty seed funding. The content of this work does not necessarily reflect the position or policy of the sponsors and no official endorsement should be inferred. The authors would like to thank Sarah Belousov, Ermine Teves, Anna Kasunic and other members of the rCommerce Laboratory at Carnegie Mellon University for their valuable contributions during the needs assessment phase. We would also like to thank our community partners, Western Pennsylvania School for Blind Children, Western Pennsylvania School for the Deaf, and Blind and Vision Rehabilitation Services of Pittsburgh for their valuable feedback during design, development, and testing phases.

References

1. Apostolopoulos, I., Fallah, N., Folmer, E., Bekris, K.E.: Feasibility of interactive localization and navigation of people with visual impairments. In: 11th IEEE Intelligent Autonomous Systems (IAS10) (2010)
2. Bernardos, A., Casar, J., Tarrio, P.: Real time calibration for rss indoor positioning systems. In: 2010 International Conference on Indoor Positioning and Indoor Navigation (IPIN), pp. 1–7, Sep (2010)
3. Cagigas, D.: Hierarchical algorithm with materialization of costs for robot path planning. Robot. Auton. Syst. **52**(2–3), 190–208 (2005)
4. Evennou, F., Marx, F.: Advanced integration of WiFi and inertial navigation systems for indoor mobile positioning. EURASIP J. Adv. Signal Process. **2006**, 1–11 (2006)
5. Feldmann, S., Kyamakya, K., Zapater, A., Lue, Z.: An indoor Bluetooth-based positioning system: concept, implementation and experimental evaluation. In: Proceedings of the International Conference on Wireless Networks, pp. 109–113 (2003)
6. Horstmann, M., Lorenz, M., Watkowski, A., Ioannidis, G., Herzog, O., King, A., Evans, D.G., Hagen, C., Schlieder, C., Burn, A.-M., King, N., Petrie, H., Dijkstra, S., Crombie, D.: Automated interpretation and accessible presentation of technical diagrams for blind people. New Rev. Hypermedia Multimedia **10**(2), 141–163 (2004)
7. Hub, A.: Precise indoor and outdoor navigation for the blind and visually impaired using augmented maps and the TANIA system. In: Proceedings of the 9th International Conference on Low Vision (2008)
8. Jin, Y., Toh, H.-S., Soh, W.-S., Wong, W.-C.: A robust dead-reckoning pedestrian tracking system with low cost sensors. In: 2011 IEEE International Conference on Pervasive Computing and Communications (PerCom), Mar (2011)

9. Korsah, G., Stentz, A., Dias, M.: Dd* lite: Efficient Incremental Search with State Dominance. Technical Report CMU-RI-TR-07-12. Robotics Institute, Pittsburgh, May (2007)
10. Kothari, N., Kannan, B., Glasgow, E.D., Dias, M.B.: Robust indoor localization on a commercial smart phone. Procedia Comput. Sci. **10**, 1114–1120. ISSN **1877–0509**, (2012). doi:10.1016/j.procs.2012.06.158
11. Ladner, R.E., Ivory, M.Y., Rao, R., Burgstahler, S., Comden, D., Hahn, S., Renzelmann, M., Krisnandi, S., Ramasamy, M., Slabosky, B., Martin, A., Lacenski, A., Olsen, S., Groce, D.: Automating tactile graphics translation. In: Proceedings of the 7th international ACM SIGACCESS conference on Computers and accessibility, Baltimore (2007)
12. Luo, X., O'Brien, W.J., Julien, C.L.: Comparative evaluation of received signal-strength index (SSI) based indoor localization techniques for construction jobsites. Adv. Eng. Inform. **25**(2), 355–363 (2011)
13. Makino, H., Ishii, I., Nakashizuka, M.: Development of navigation system for the blind using GPS and mobile phone combination. In: 18th Annual International Conference of the IEEE Engineering in Medicine and Biology Society (1996)
14. Mau, S., Melchior, N.A., Makatchev, M., Steinfeld, A.: BlindAid: An Electronic Travel Aid for the Blind. Robotics Institute, Report, CMU-RI-TR-07-39 (2008)
15. Oktem, R., Aydin, E.: An RFID based indoor tracking method for navigating visually impaired people. Turk. J. Electr. Eng. Comput. Sci. **18**(2), 185–195 (2010)
16. Ran, L., Helal, S., Moore, S.: Drishti: An Integrated Indoor/Outdoor Blind Navigation System and Service. In: Proceedings of the 2nd IEEE Annual Conference on Pervasive Computing and Communications (2004)
17. Sanchez, J.: Mobile Audio Navigation Interfaces for the Blind. Universal Access in HCI, Part II, HCII 2009, LNCS 5616, pp. 236–245, (2009)
18. Wang, H., Lenz, H., Szabo, A., Bamberger, J., Hanebeck, U.: WLAN-based pedestrian tracking using particle filters and low-cost mems sensors. In: 4th Workshop on Positioning, Navigation and Communication, Mar (2007)
19. Wendlandt, K., Berhig, M., Robertson, P.: Indoor localization with probability density functions based on bluetooth. In: 16th IEEE International Symposium on Personal, Indoor and Mobile Radio Communications (2005)

Chapter 4
Real-Time Modeling of Ocean Currents for Navigating Underwater Glider Sensing Networks

Dongsik Chang, Xiaolin Liang, Wencen Wu, Catherine R. Edwards and Fumin Zhang

Abstract Ocean models that are able to provide accurate and real-time prediction of ocean currents will improve the performance of glider navigation. In this paper, we propose a novel approach to compute a model for ocean currents at higher resolution than existing approaches. By focusing on a small area and incorporating measurements from multiple gliders, we are able to perform real-time computation of the model, which can be used to improve performance of underwater glider navigation in the ocean. Our model uses a lower resolution, larger scale dataset generated from existing models to initialize the computation. We have also demonstrated incorporating data streams from high frequency (HF) radar measurements of surface currents. Glider navigation performance using the proposed ocean currents model is demonstrated in a simulated flow field based on data collected off the coast of Georgia, USA.

The research work is supported by ONR grants N00014-08-1-1007, N00014-09-1- 1074, and N00014-10-10712 (YIP), and NSF grants ECCS-0841195 (CAREER), CNS-0931576, ECCS-1056253, and OCE-1032285.

D. Chang (✉), X. Liang, W. Wu and F. Zhang
School of Electrical and Computer Engineering, Georgia Institute of Technology,
Atlanta, GA 30332, USA
e-mail: dsfrancis3@gatech.edu

X. Liang
e-mail: xliang31@gatech.edu

W. Wu
e-mail: wwencen3@gatech.edu

F. Zhang
e-mail: fzhang@gatech.edu

C. R. Edwards
Skidaway Institute of Oceanography, Savannah, GA 31411, USA
e-mail: catherine.edwards@skio.usg.edu

A. Koubâa and A. Khelil (eds.), *Cooperative Robots and Sensor Networks*,
Studies in Computational Intelligence 507, DOI: 10.1007/978-3-642-39301-3_4,

Keywords Mobile sensing network · Underwater glider navigation · Ocean circulation model

1 Introduction

Ocean circulation models are computational models used to study ocean processes that are employed for ocean flow predictions. Such models include the Regional Ocean Modeling System (ROMS) [1], the Terrain-following Ocean Modeling System (TOMS) [2], and the Hybrid Coordinate Ocean Model (HYCOM) [3], just to name a few. They have broad applications in navigation [4–6].

Underwater gliders are moving robotic sensing platforms [7] that are able to perform persistent surveying missions in the ocean to collect data that can be used to significantly improve the accuracy of ocean circulation models [8, 9]. Depth-averaged ocean current can be estimated from the GPS fixes obtained when a glider surfaces [10]. However, these direct measurements of the current can only provide limited help for glider navigation because most existing path planning algorithms [11, 12] require the knowledge of a distribution of the flow, e.g., 4-D flow field measured or predicted by ocean models. These models contain errors caused by limited model resolution, approximations in mathematical modeling (e.g., incomplete physics such as parameterization of turbulence closure), and numerical error in computation, and have a low spatial and temporal resolution relative to the horizontal scale of glider motion. While oceanic data collected by gliders is useful for assimilation into real-time models, the data cannot be immediately assimilated because of limited data communication speed and the significant amount of computation time for data assimilation for the ocean model, causing time delays for path planning [13].

We propose to develop a high-resolution ocean model that provides real-time predictions of the depth-averaged ocean currents in a relatively small region near the glider. Our model, designed to enable efficient path planning, is initialized using data from a lower resolution ocean model. After the initialization, the model will be updated only based on the measurements from the glider sensing network. We use spatial and temporal basis functions and their corresponding coefficients to approximate the ocean flow field. We model ocean currents as two decomposed components: tidal and non-tidal components. We choose sinusoidal functions with deterministic frequencies as basis functions for the tidal component and weighted Laguerre polynomials as temporal basis functions for the non-tidal component. The frequencies of sinusodial functions match the natural frequencies of the tidal constituents that are dominant in a specific area. The Gaussian radial basis functions (RBFs) are chosen as the spatial basis functions to represent the flow field. Measurements of current are collected by a group of gliders that remain in a formation under control laws based on the Jacobi transform [14]. In practice [8], the gliders can communicate with an onshore computer to report their measurements when they surface. The computer will use the measurements to recompute the coefficients of the basis functions in our model so that a real-time model of the ocean current can be generated. Using this model, control commands to navigate the gliders are then generated and delivered. This work extends our previous publication [15], which models only tidal current using sinusoidal basis functions, by incorporating the modeling of a non-tidal current component, hence, provides a more general approach for ocean current modeling.

We organize the remainder of this paper in the following way: Sect. 2 provides the model setup using basis functions, Sect. 3 describes the initialization process of the model, Sect. 4 develops the techniques used to update the model in real time, Sect. 5 presents the simulation results, and Sect. 6 provides the conclusion.

2 Representation of Ocean Currents

Ocean currents have both temporal and spatial variations, which we propose to approximate with a series of temporal and spatial basis functions. To model the temporal characteristics, we decompose ocean currents into a tidal component and a non-tidal component since tidal currents are forced by a superposition of known tidal constituents, each of which has specific frequency related to astronomical phenomena. We model temporal variation of the tidal component by using a series of sinusoidal functions as the temporal basis functions with frequencies of the tidal constituents. In oceanographic research, the symbol u denotes flow velocity in E/W direction and v denotes flow velocity in N/S direction. We use ξ to denote either u or v. Suppose we want to estimate tidal currents around a glider at position $\mathbf{x}$. Then, the tidal component ξ is expressed as

$$\xi_{\text{tide}}(\mathbf{x}, t) = \bar{\xi}_{\text{tide}}(\mathbf{x}) + \sum_{i=1}^{N} [g_i(\mathbf{x}) \cos(\omega_i t) + h_i(\mathbf{x}) \sin(\omega_i t)], \tag{1}$$

where N is the number of tidal constituents in the area and $\bar{\xi}_{\text{tide}}(\mathbf{x})$ is a tidal residual. $\bar{\xi}_{\text{tide}}(\mathbf{x})$, $g_i(\mathbf{x})$, and $h_i(\mathbf{x})$ are coefficients corresponding to temporal basis functions. Given the position of a glider $\mathbf{x}$, $\bar{\xi}_{\text{tide}}(\mathbf{x})$, $g_i(\mathbf{x})$, and $h_i(\mathbf{x})$ can be approximated by a series of spatial functions.

It is known that radial basis functions (RBFs) have less approximation error than polynomials [16]. Gaussian RBFs have better trade-off between accuracy and smoothness of the approximation than other radial basis functions [17]. Therefore, we use Gaussian RBFs as the spatial basis functions. Then, $\bar{\xi}_{\text{tide}}(\mathbf{x})$, $g_i(\mathbf{x})$, and $h_i(\mathbf{x})$ are

$$\bar{\xi}_{\text{tide}}(\mathbf{x}) = \sum_{j=1}^{M} \kappa_{1j} \Phi_j(\mathbf{x}),$$
$$g_i(\mathbf{x}) = \sum_{j=1}^{M} \kappa_{2i,j} \Phi_j(\mathbf{x}), \tag{2}$$
$$h_i(\mathbf{x}) = \sum_{j=1}^{M} \kappa_{2i+1,j} \Phi_j(\mathbf{x}),$$

where M is the number of RBFs used, $\kappa_{i,j}$ are the coefficients of the Gaussian RBFs, $\Phi_j(x) = \exp(\frac{-\|\mathbf{x}-\mathbf{c}_j\|^2}{2\sigma^2})$, c_j's are the centers, σ is the width. As $\bar{\xi}_{\text{tide}}(\mathbf{x})$, $g_i(\mathbf{x})$, and $h_i(\mathbf{x})$ are functions of the position of the glider, they share the same centers and width of Gaussian RBFs. This is an empirical model, and unlike a physical model, it may not account for the nature of known and unknown ocean dynamics.

To approximate non-tidal currents, weighted Laguerre polynomials are chosen as temporal basis functions. The weighted Laguerre polynomials are orthogonal and force temporal components decay exponentially as $t \to \infty$, maintaining the system stable. The rate of exponential decay is determined by the time scaling factor. The $n-$th order weighted Laguerre polynomials are defined as

$$L_n(t) = \sqrt{2p}\frac{e^{pt}}{n!}\frac{d^n}{dt^n}[t^n e^{-2pt}], \tag{3}$$

where $n = 0, 1, 2, \cdots$ and p is the time scaling factor, which adjusts the decaying rate of the weight. Using 0- to P-th order Laguerre polynomials, we model non-tidal currents as

$$\xi_{\text{nontide}}(\mathbf{x}, t) = \sum_{n=0}^{P} q_n(\mathbf{x}) L_n(t), \tag{4}$$

where P is the order of the polynomial functions and q_n is defined as

$$q_n(\mathbf{x}) = \sum_{j=1}^{M} \mu_{n,j} \Phi_j(\mathbf{x}), \tag{5}$$

where M is the number of RBFs used and $\mu_{n,j}$ are coefficients of the RBFs Φ_j's. For the same location $\mathbf{x}$, ξ_{nontide} uses the same parameters c, σ of the RBFs as in ξ_{tide}. Then, ocean currents are defined as the summation of tidal currents and non-tidal currents as

$$\xi(\mathbf{x}, t) = \xi_{\text{tide}}(\mathbf{x}, t) + \xi_{\text{nontide}}(\mathbf{x}, t). \tag{6}$$

Figure 1 describes the structure of the ocean currents model. The model takes the position $\mathbf{x}$ as input and computes RBFs Φ. Then, by summing up all the weighted

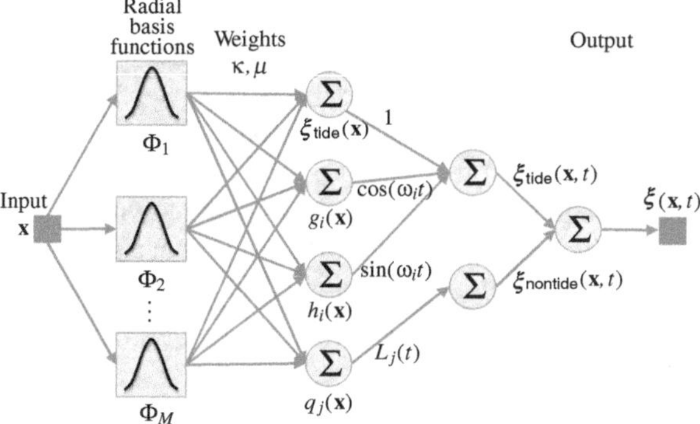

Fig. 1 Structure of the ocean currents model. The model takes a position as input and generates radial basis functions (RBF) with respect to the position. Then, RBFs and their weights form coefficients corresponding to each element of temporal basis functions. Weighted sinusoidal functions and polynomial functions using their corresponding coefficients form tidal currents and non-tidal currents, respectively. Then, summation of these two currents produces the ocean currents

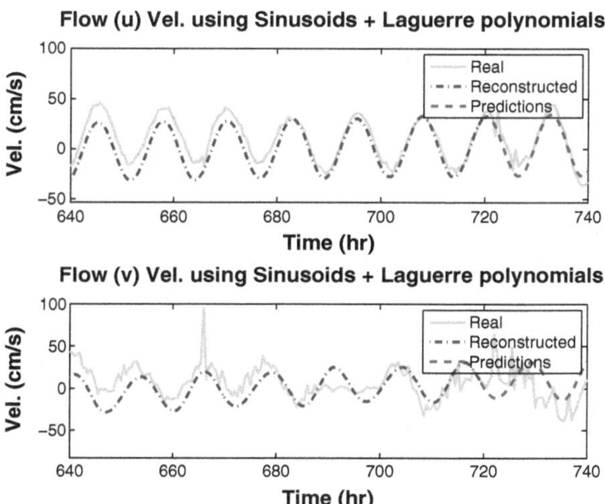

Fig. 2 Ocean currents at coordinate (−79.8984,31.3531) modeled by sinusoidal and Laguerre polynomial basis functions: Real flow (*light blue line*), reconstructed flow (*blue dashdotted line*), and predicted flow (*red dashed line*). Weights $\kappa_{i,j}$, $\mu_{n,j}$ for the spatial basis functions are generated using 30-day historical data from January 1, 2011 to reconstruct the flow and predictions are produced for the next 20 h without updating $\kappa_{i,j}$, $\mu_{n,j}$. For detiding, we use T_Tide MATLAB® toolbox [18]. T_Tide extracts 29 constituents from the 30-day record, with the M2 tide (approximately 12.42 h period) as the most significant

RBFs using $\kappa_{i,j}$ or $\mu_{n,j}$, coefficients for each temporal basis function are generated. Each linear combination of sinusoidal functions and polynomials weighted by corresponding coefficients forms tidal currents ξ_{tide} and non-tidal currents ξ_{nontide}. Figure 2 shows observed ocean flow and reconstructed flow by summing up ξ_{tide} and

ξ_{nontide}, which are generated from the observed flow as the input to the model. The figure also demonstrates a comparison between the observation and predicted flow using the model.

Data fitting and model reconstruction, or approximation, by our method using RBF may seem similar to Gaussian process regression. The former uses a deterministic model to interpolate a target value based on given data at known locations, whereas the latter uses a stochastic model to estimate data values at a particular location based on known covariance of the field. The relationship between RBF interpolation and Gaussian process regression is discussed in [19].

3 Model Initialization

We use data from a lower-resolution ocean model to initialize the computation of our model. To construct a smaller-scale, higher-resolution ocean currents model, only the ocean flow in the vicinity of the glider is considered. First, we select a $n \times n$ grid in the existing ocean model and refer to it as a *patch*, which is illustrated as red and yellow boxes in Fig. 3. The initial model is built by interpolating data from the grid points in the patch according to Eqs. (1), (4), and (6).

The number of RBF centers M should be smaller than the number of grid points n^2 [20]. Instead of choosing random positions in the patch as centers, we use K-means clustering method [21] to compute the positions of the centers. The K-means clustering algorithm seeks to partition the grid points $\mathbf{x}_i, i = 1, \cdots, n^2$, into M disjoint subsets S_j, each containing n_j data points in such a way as to minimize $J = \sum_{j=1}^{M} \sum_{\mathbf{x}_i \in S_j} \|\mathbf{x}_i - \mathbf{c}_j\|^2$. The width σ of the RBFs will affect the smoothness of the interpolation. One commonly used strategy is to choose σ to be the same spatial scale with the input data [16]; in our case σ, is proportional to the grid size (in meters).

Given the basis functions and the time-series data of the grid points, we can use the least mean square method to compute the coefficients $\kappa_{i,j}, \mu_{n,j}$ to generate the initial

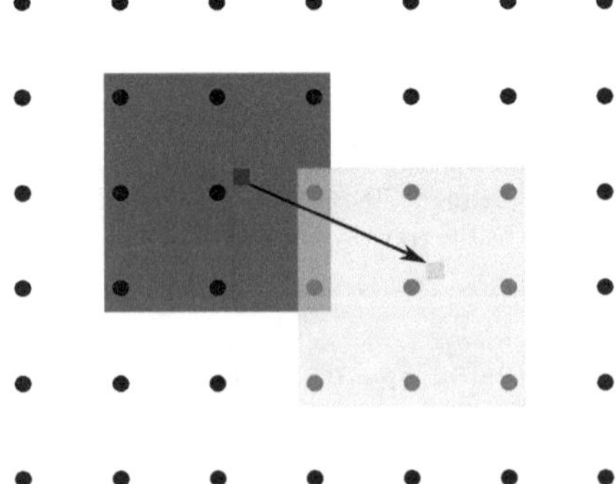

Fig. 3 Grid points used by ocean models. In this figure the *black round dots* represents the grid points where data from larger scale ocean models are available to initialize our model. The *blue rectangular dot* represents the initial position of the formation center

model. With this model, we can approximate the flow velocity given the position $\mathbf{x}$ of a glider. A group of gliders in a sensing network will be controlled to maintain a formation and navigate in the area, as illustrated by the red patch in Fig. 3. The center of the formation is represented by the blue square dot. However, when they move out of the covered area, e.g., the red patch, the RBFs will have difficulty in representing the flow near the gliders' positions outside the previous patch, e.g., the yellow patch in Fig. 3. Therefore, we need to re-initialize the RBFs' centers based on the current positions of the gliders. We update the positions of the centers by reapplying K-means clusteringing method once the gliders move out of the covered area by the RBFs. The change of centers leads to the change of the RBFs.

4 Flow Field Map Update

To incorporate most recent ocean dynamics, we need to update weights of spatial basis functions $\kappa_{i,j}$, $\mu_{n,j}$. Plugging Eqs. (2) and (5) into Eqs. (1) and (4), we obtain the model of ocean currents as

$$\begin{aligned}\xi_{\text{tide}}(\mathbf{x},t) &= \sum_{j=1}^{M}\kappa_{1j}\Phi_j(\mathbf{x}) + \sum_{i=1}^{N}\left[\sum_{j=1}^{M}\kappa_{2i,j}\Phi_j(\mathbf{x})\cos(\omega_i t)\right.\\ &\quad \left. + \sum_{j=1}^{M}\kappa_{2i+1,j}\Phi_j(\mathbf{x})\sin(\omega_i t)\right]\\ &= \sum_{j=1}^{M}\left(\kappa_{1,j} + \sum_{i=1}^{N}\kappa_{2i,j}\cos(\omega_i t) + \sum_{i=1}^{N}\kappa_{2i+1,j}\sin(\omega_i t)\right)\Phi_j(\mathbf{x})\\ &= \sum_{j=1}^{M}\alpha_j(t)\Phi_j(\mathbf{x}), \end{aligned} \tag{7}$$

and

$$\begin{aligned}\xi_{\text{nontide}}(\mathbf{x},t) &= \sum_{n=0}^{P}\sum_{j=1}^{M}\mu_{n,j}\Phi_j(\mathbf{x})L_n(t)\\ &= \sum_{j=1}^{M}\sum_{n=0}^{P}\mu_{n,j}L_n(t)\Phi_j(\mathbf{x})\\ &= \sum_{j=1}^{M}\beta_j(t)\Phi_j(\mathbf{x}), \end{aligned} \tag{8}$$

where

$$\begin{aligned}\alpha_j(t) &= \kappa_{1,j} + \sum_{i=1}^{N} \kappa_{2i,j} \cos(\omega_i t) + \sum_{i=1}^{N} \kappa_{2i+1,j} \sin(\omega_i t),\\ \beta_j(t) &= \sum_{n=0}^{P} \mu_{n,j} L_n(t).\end{aligned} \tag{9}$$

Define $\alpha = [\alpha_1, \dots, \alpha_M]^T$ and $\beta = [\beta_1, \dots, \beta_M]^T$. We stack $\alpha(t)$ and $\beta(t)$, and define $\xi(t) = [\alpha(t), \beta(t)]^T$.

Recent flow velocity estimates derived from glider on-board sensors will be used to update the parameters of the ocean currents model. We want to find $\kappa_{i,j}$ and $\mu_{n,j}$ so that the mean square error between the predicted ocean flow $\hat{\xi}(\mathbf{x}, t)$ and the true flow $\xi(\mathbf{x}, t)$ is minimized. It is well known that Kalman filter provides a solution [22]. Assume there are $2M$ gliders obtaining the measurements simultaneously. (The choice of M will be explained later in this section.) At time step k, the measurements from the gliders are denoted by $z(k) = [z_1(k), z_2(k), \dots, z_{2M}(k)]^T$. Then, we have the measurement equation of the Kalman filter,

$$z(k) = H(k)\xi_k + v(k), \tag{10}$$

where $H(k)$ is a matrix with the i-th row defined as Gaussian radial basis functions $[\Phi_1(\mathbf{x}_i), \dots, \Phi_M(\mathbf{x}_i), \Phi_1(\mathbf{x}_i), \dots, \Phi_M(\mathbf{x}_i)]$, $i = 1, \dots, M$. $\Phi_j(\mathbf{x})$, $j = 1, \dots, M$ are Gaussian spatial basis functions. The measurement noise vector $v(k) = [v_1(k), v_2(k), \dots, v_{2M}(k)]^T$ is assumed to be Gaussian with zero mean and known covariance matrix R. If the properties of the noises are unknown, we can still retrieve the estimation by using an H_∞ filter [22].

According to Eqs. (7) and (8), given the instant measurements $z(k)$, we need to estimate the state ξ_k to update the ocean currents model. Suppose the sampling rate is T_s. The dynamics of the state is required to run the Kalman filter. At time $t = t_k$, Eq. (9) becomes:

$$\begin{aligned}\alpha_j(kT_s) &= \kappa_{1,j} + \sum_{i=1}^{N} \kappa_{2i,j} \cos(\omega_i kT_s) + \sum_{i=1}^{N} \kappa_{2i+1,j} \sin(\omega_i kT_s),\\ \beta_j(kT_s) &= \sum_{n=0}^{P} \mu_{n,j} L_n(kT_s).\end{aligned} \tag{11}$$

We assume that the sampling frequency is higher enough than the frequency of the tidal constituents so that it is sufficient to use Taylor series up to first order to estimate $\alpha_j((k-1)T_s)$ and $\beta_j((k-1)T_s)$. Then, using Taylor expansion, we have

$$\alpha_j(kT_s) = \alpha_j((k-1)T_s) + \sum_{i=1}^{N} \omega_i T_s \left[\kappa_{2i,j} \; \kappa_{2i+1,j} \right] \begin{bmatrix} -\sin(\omega_i(k-1)T_s) \\ \cos(\omega_i(k-1)T_s) \end{bmatrix},$$
$$\beta_j(kT_s) = \beta_j((k-1)T_s) + \sum_{n=0}^{P} \mu_{n,j} L'_n((k-1)T_s), \tag{12}$$

where L'_n is the first order derivative of each polynomial function L_n. We define $f_{\alpha,i}(k-1) = \omega_i T_s \begin{bmatrix} \kappa_{2i,1} & \kappa_{2i+1,1} \\ \vdots & \vdots \\ \kappa_{2i,M} & \kappa_{2i+1,M} \end{bmatrix} \begin{bmatrix} -\sin(\omega_i(k-1)T_s) \\ \cos(\omega_i(k-1)T_s) \end{bmatrix}$ and $f_{\beta,n}(k-1) = \mu_{n,j} L'_n((k-1)T_s)$. Let us denote $\xi_k = \xi(kT_s) = [\alpha_1(kT_s), \ldots, \alpha_M(kT_s), \beta_1(kT_s), \ldots, \beta_M(kT_s)]^T$. Now, we have the state equation for the Kalman filter:

$$\xi_k = \xi_{k-1} + \begin{bmatrix} \sum_{i=1}^{N} f_{\alpha,i}(k-1) \\ \sum_{n=0}^{P} f_{\beta,n}(k-1) \end{bmatrix} + w(k). \tag{13}$$

Here, we introduce $w(k) = [w_{\alpha,1}(k), \ldots, w_{\alpha,M}(k), w_{\beta,1}(k), \ldots, w_{\beta,M}(k)]^T$ as the process noise at the $k-$th step. We assume w_k is Gaussian with zero mean and known covariance $Q(k)$. Define $\hat{\xi}_k$ to be the optimal estimate of ξ_k and $P(k)$ the $2M \times 2M$ covariance matrix corresponding to the uncertainty of the estimate. Then, a standard Kalman filter can be computed based on the state Eq. (13) and measurement Eq. (10).

Given updated $\hat{\xi}_j(k) = [\hat{\alpha}_j(k), \hat{\beta}_j(k)]^T$, $j = 1, \ldots, M$, we are able to use the recursive least mean square (RLMS) method to update $\kappa_{i,j}$ and $\mu_{n,j}$ in Eq. (11). For $\kappa_{i,j}$, define $\psi(k) = [1, \cos(\omega_1 kT_s), \sin(\omega_1 kT_s), \ldots, \cos(\omega_N kT_s), \sin(\omega_N kT_s)]$, and $\kappa_j = [\kappa_{1,j}, \kappa_{2,j}, \ldots, \kappa_{2i+1,j}]^T$. Then, we have $\hat{\alpha}_j(k) = \psi(k)\kappa_j$. According to RLMS method, we can derive the following equations:

$$K(k) = \Sigma(k-1)\psi^T(k)(\Gamma(k) + \psi(k)\Sigma(k-1)\psi^T(k))^{-1} \tag{14}$$
$$\hat{\kappa}_j(k) = \hat{\kappa}_j(k-1) + K(k)[\hat{\alpha}_j(k) - \psi(k)\hat{\kappa}_j(k-1)] \tag{15}$$
$$\Sigma(k) = (I - K(k)\psi(k))\Sigma(k-1), \tag{16}$$

where $\Sigma(k)$ is the error covariance matrix, $K(k)$ is the gain, and $\Gamma(k)$ is the noise covariance at time step k. Similarly, we can update $\mu_{n,j}$ using RLMS. Plugging the updated $\kappa_{i,j}$ and $\mu_{n,j}$ into the ocean currents model, we are able to provide predictions of the flow velocity within the grid size of a glider.

However, when the gliders move out of the original patch, to provide more accurate estimation of the tidal flow around the gliders, RBFs are changed as discussed in Sect. 3. Therefore, the position matrix H will change, which leads to the change of the measurement equation in Eq. (10) of the Kalman filter. As a patch change can happen before the Kalman filter converges to a steady state, reinitializing the Kalman filter might result in unreliable estimation. Assume the gliders move out of the original patch at time step k^*, the new position matrix is $\tilde{H}$, and the new state is

$\tilde{\xi}$. Then, in the area where the previous and current patches overlap, we have $z(k^*) = \tilde{H}(k^*)\tilde{\xi}_{k^*} = H(k^*)\xi_{k^*}$. According to Eq. (10), if $\tilde{H}(k^*)$ is invertible, i.e., $\tilde{H}(k^*)$ is a square non-singular matrix, then $\tilde{\xi}_{k^*}$ can be solved by $\tilde{\xi}_{k^*} = \tilde{H}^{-1}(k^*)H(k^*)\xi_{k^*}$.

To ensure that H is invertible, we need to apply the following rule. With M RBFs used to approximate the ocean flow, H consists of $2M$ columns. Since each row vector of H corresponds to the list of RBFs for α_j, β_j, $j = 1, 2, \ldots, M$ used by one glider, the position matrix H should also have $2M$ rows. That is, $2M$ gliders need to take measurements at the same time to ensure that H is a square matrix. This fact actually provides a relationship between the number of gliders used and the number of RBFs. If the number of gliders are less than two times the number of RBFs, then the size of the patch should be selected to make sure the Kalman filter converges before the re-initialization. In this case, the patch cannot be very small, which makes our model hard to represent the flow around the gliders. If the number of gliders equals two times of the number of RBFs selected, then such limitation on the patch size does not have to be considered, and the patch can be selected smaller. Hence, the flow field may be generated with higher accuracy. However, in scenarios that we only care one component in the ocean current (tidal component or nontidal component), the number of gliders can be reduced to M since the state $\xi(t)$ only consists of $\alpha(t)$ or $\beta(t)$. For example, in areas that tidal current are dominant, we can model the ocean current only using the tidal component.

5 Simulation Results

We use a near real-time data stream of high frequency (HF) radar measurements of surface current from a long range Wave Radar (WERA) system [23, 24], operated and maintained by Dana Savidge at the Skidaway Institute of Oceanography, to evaluate the feasibility of our model. The WERA system is a shore-based remote sensing system using the over-the-horizon radar technology to monitor ocean surface currents. Shore stations of HF radar emit radio signals that bounce off surface waves and return to receiver. The received radio wave is used to compute ocean surface current movement relative to ocean surface wave movement. Generally, hourly data with 6 km grid spacing are published online with a 3 h processing delay, but in our simulation, we use data of $3 \times 3\,\text{km}^2$ spatial resolution and 30 min temporal resolution.

The Gulf Stream is the dominant feature in ocean currents along the east coast of the United States. Near the edge of the continental shelf, most of the low-frequency (period > 2 day) part is accounted for by the Gulf Stream, but its influence still remains in the region where tidal currents (period $<$ 26 h) are dominant and low frequency variability is due to wind-driven currents. We detide sea surface currents data using T_Tide MATLAB® toolbox [18]. We use 30-day historical data from January 1, 2011 to initialize our model. For the data we used, T_Tide extracts total 29 constituents including M2 and N2 as two major ones, and all the $N = 29$ constituents are used to generate weight parameters for the tidal component in the model. For the non-tidal component, we use 0–$P = 10$th order Laguerre polynomials.

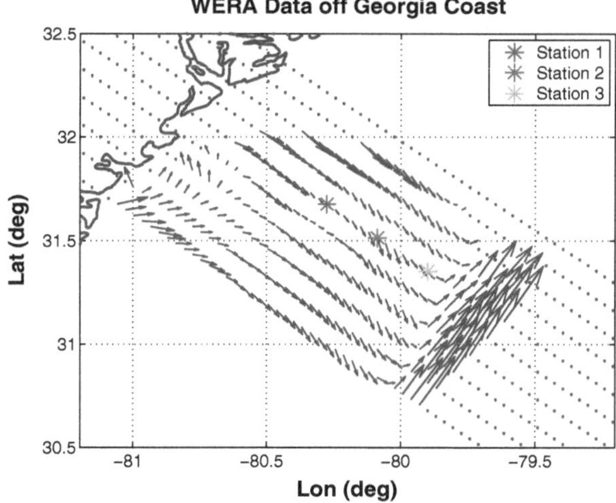

Fig. 4 WERA data off Georgia Coast courtesy of Dana Savidge at the Skidaway Institute of Oceanography. Data have $3 \times 3\,\text{km}^2$ spatial resolution and 30 min temporal resolution. The Gulf Stream featured by strong Northeast-ward flow can be identified in the figure. For each simulation, we deploy a group of 6 gliders near one of the 3 different locations illustrated as *colored stars*. The initial positions of the gliders are selected so that the center of the group coincides with the starred location. The group will then be asked to maintain its center location by canceling estimated flow. The coordinates of the starred locations are (−80.2752,31.6771), (−80.0868,31.5151), (−79.8984,31.3531) (from station 1 to 3)

We show three experiments using historical WERA data to generate a simulated flow field. For each simulation, we simulate a group of 6 gliders that are deployed near one of the starred locations. The ocean current model is initialized by using data from a 5×5 patch around the starred location to compute the center and the weights of the basis functions. We select three centers for the RBFs in the patch, hence 6 gliders are desired. Using the positions of the grid points in the selected patch and the K-means clustering method, we can obtain the positions of the 3 centers $[\mathbf{c}_1, \mathbf{c}_2, \mathbf{c}_3]$ for the RBF. When the center of the glider formation reaches 5 km away from the center of each patch, we reinitialize the patch by recomputing the centers and the weights of basis functions. The σ parameter for the RBF is selected to be 3000 m.

We initialize our model using a 30-day WERA data segment from January 1, 2011. After the model is initialized, we select another segment of data from January 31, 2011 to represent the real ocean. We use linear interpolation to generate flow values at the location of the gliders. Then, we simulate glider flow measurements by adding measurement noise to the flow value at the location of the glider. This setup resembles the situation where the gliders are moving in the real ocean, and the only information we have about the ocean is the measurements from the gliders.

To compare with the proposed method, we provide simulation results using the Advanced Circulation (ADCIRC) predictive ocean model [25], which solves the generalized wave continuity equation on triangular finite elements, and has been widely

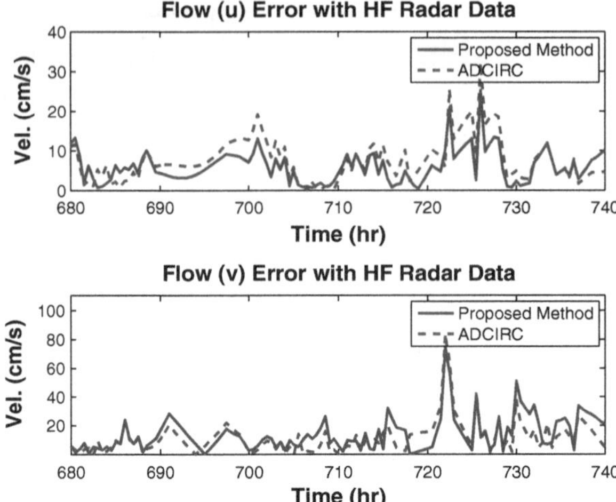

Fig. 5 Error comparison of our model and the ADCIRC predictive ocean model with HF radar data at coordinate (−79.8984,31.3531): Our model (*blue line*) and ADCIRC (*green dashed line*). Six tidal constituents (M2, S2, N2, K1, and O1) that have a largest influence off Georgia, USA are used to reconstruct the tidal current, and the element-wise Euclidean norm is used as an error metric. Unlike ADCIRC, our model estimates other ocean current components as well as tidal current, and it shows better performance here since the testing location is under the influence of the Gulf Stream

used for studying tidal and wind-driven flow, storm surge, and other circulation and transport problems. The ADCIRC model implementation used here [26] is a tidal database created based on simulation runs using ADCIRC. Tidal current generated using ADCIRC is compared with our proposed method in Fig. 5, in which model errors are evaluated using the element-wise Euclidean norm.

After being deployed, the gliders will attempt to maintain their initial position by canceling the estimated flow generated by our ocean model. We assume that a glider can cancel an estimated flow. This will result in its drifting under actual flow due to error in the estimated flow. Assuming the formation of the glider is maintained, the motion of the center of a glider fleet is expressed as $\mathbf{x}(k+1) = \mathbf{x}(k) - \hat{\xi}(\mathbf{x}, k)T_s + \xi(\mathbf{x}, k)T_s$, where $\hat{\xi}(\mathbf{x}, k)$ is the estimated flow at position $\mathbf{x}$ and time t and $\xi(\mathbf{x}, k)$ is the real flow.

In each of the three simulation, the navigation performance is evaluated by measuring the distance between the center of the glider formation and the corresponding starred location in Fig. 4. The coordinates of the starred location are (−80.2752,31.6771), (−80.0868,31.5151), (−79.8984,31.3531) (from station 1 to 3). Incorporating the measurements taken by the gliders into the Kalman filter, we are able to update the ocean currents model in real time. For the Kalman filter, the measurement noise variance of each glider is $R = 0.01I_6$, the process noise covariance is $Q = 0.01I_6$, and initial error covariance matrix is $P(0) = I_6$. For RLMS,

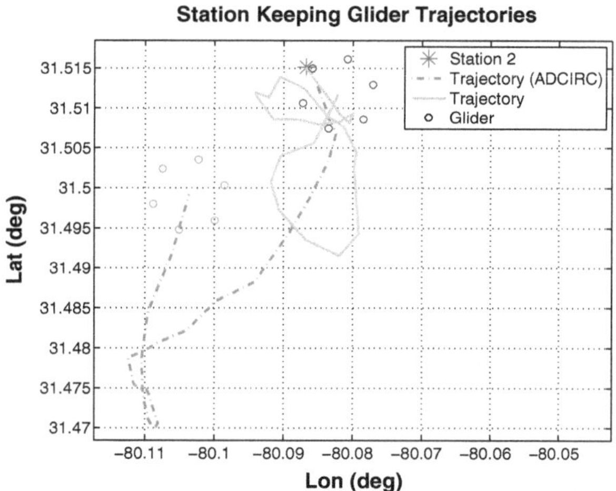

Fig. 6 The trajectories of the center of a station keeping glider fleet under our model (*light blue line*) and under ADCIRC (*orange dashdotted line*). Gliders attempt to maintain their positions by heading towards the opposite direction of the estimated flow

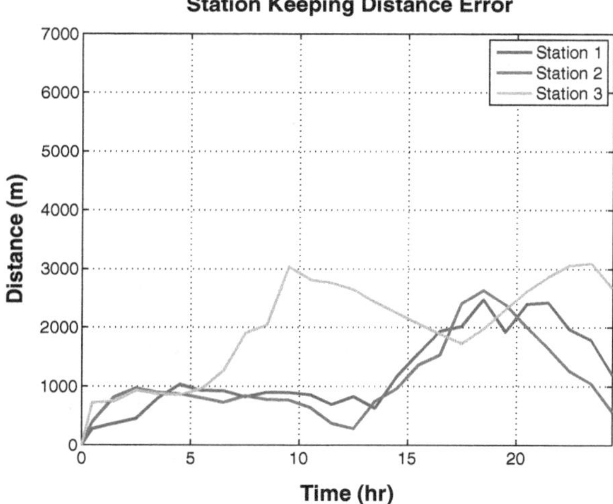

Fig. 7 Distance between the center of the groups of 6 gliders and the starred locations in a 24 h period when flow canceling controller is used

the noise covariance is $\Gamma = 0.01$ and the error covariances for the tidal and non-tidal components are $\Sigma_{\text{tide}} = 0.01 I_{2N+1}$ and $\Sigma_{\text{nontide}} = 0.01 I_{P+1}$, respectively.

Simulation results show that the formation attempts to maintain their positions and stays closer to the target station under our model, see Fig. 6. Due to error in flow estimate, the distance between the formation centers and the starred locations grows gradually as shown in Fig. 7. Note that our controller only cancels the estimated flow, but does not navigate gliders towards the starred position. Even with this simple controller, navigation error using our model, i.e., approximately 3000 m in 24 h, is a significant improvement over results when lower resolution flow model was used as in [27, 5].

6 Conclusion

Accurate prediction of ocean currents is essential in underwater glider navigation. In this paper, we combine data in existing ocean models with instant measurements taken by a group of gliders in a sensing network to build a high-resolution real-time current model. This model generates distribution of ocean flow around the gliders for underwater path planning. The update of the model only requires instant measurements, thus, is computationally efficient compared with the data assimilation procedure in the existing ocean models. Further work will be explored to use a group of gliders in a sensing network working cooperatively to build ocean model in 3-D environment and depth variation of the flow velocity will be addressed therein.

References

1. Shchepetkin, A.F., McWilliams, J.C.: The regional oceanic modeling system (ROMS): a split-explicit, free-surface, topography-following-coordinate oceanic model. Ocean Model. **9**(4), 347–404 (2005)
2. Haidvogel, D.B., Arango, H., Budgell, W.P., Cornuelle, B.D., Curchitser, E., Di Lorenzo, E., Fennel, K., Geyer, W.R., Hermann, A.J., Lanerolle, L.: Ocean forecasting in terrain-following coordinates: formulation and skill assessment of the regional ocean modeling system. J. Comput. Phys. **227**(7), 3595–3624 (2008)
3. Chassignet, E.P., Hurlburt, H.E., Smedstad, O.M., Halliwell, G.R., Hogan, P.J., Wallcraft, A.J., Baraille, R., Bleck, R.: The HYCOM (HYbrid Coordinate Ocean Model) data assimilative system. J. Mar. Syst. **65**(1–4), 60–83 (2007)
4. Smith, R.N., Chao, Y., Li, P.P., Caron, D.A., Jones, B.H., Sukhatme, G.S.: Planning and implementing trajectories for autonomous underwater vehicles to track evolving ocean processes based on predictions from a regional ocean model. Int. J. Robot. Res. **29**(12), 1475–1497 (2010)
5. Szwaykowska, K., Zhang, F.: A lower bound on navigation error for marine robots guided by ocean circulation models. In: Proceedings of IEEE/RSJ International Conference on Intelligent Robots and Systems (IROS 2011), pp. 3583–3588. San Francisco (2011)
6. Moon, C.B., Chung, W.: Design of navigation behaviors and the selection framework with generalized stochastic petri nets toward dependable navigation of a mobile robot. In: Proceedings of 2010 IEEE Conference on Robotics and Automation (May 2010), pp. 2989–2994. Anchorage (2010)
7. Lynch, K., Schwartz, I., Yang, P., Freeman, R.: Decentralized environmental modeling by mobile sensor networks. IEEE Trans. Rob. **24**(3), 710–724 (2008)
8. Zhang, F., Fratantoni, D.M., Paley, D., Lund, J., Leonard, N.E.: Control of coordinated patterns for ocean sampling. Int. J. Control. **80**(7), 1186–1199 (2007)
9. Leonard, N.E., Paley, D.A., Davis, R.E., Fratantoni, D.M., Lekien, F., Zhang, F.: Coordinated control of an underwater glider fleet in an adaptive ocean sampling field experiment in Monterey Bay. J. Field Robot. **27**(6), 718–740 (2010)
10. Bingham, B.: Predicting the navigation performance of underwater vehicles. In: Proceedings of IEEE/RSJ International Conference on Intelligent Robots and Systems (IROS 2009), pp. 261–266. St. Louis (2009)
11. Smith, R.N., Schwager, M., Smith, S.L., Jones, B.H., Rus, D., Sukhatme, G.S.: Persistent ocean monitoring with underwater gliders: adapting sampling resolution. J. Field Robot. **28**(5), 714–741 (2011)

12. Hover, F.: Path planning for data assimilation in mobile environmental monitoring systems. In: Proceedings of 2009 IEEE/RSJ International Conference on Intelligent Robots and Systems (IROS 2009), pp. 213–218. St. Louis (2009) Oct 2009
13. Paduan, J.D., Shulman, I.: HF radar data assimilation in the Monterey Bay area. J. Geophys. Res. **109**, C07S09 (2004)
14. Zhang, F., Leonard, N.E.: Cooperative control and filtering for cooperative exploration. IEEE Trans. Autom. Control. **55**(3), 650–663 (2010)
15. Liang, X., Wu, W., Chang, D., Zhang, F.: Real-time modelling of tidal current for navigating underwater glider sensing networks. Procedia Comput. Sci. **10**, 1121–1126 (2012)
16. Kumar, S.: Neural Networks: A Classroom Approach. Tata McGraw-Hill, Education, India (2004)
17. Bishop, C.M.: Neural Networks for Pattern Recognition. Oxford University Press, New York (1996)
18. Pawlowicz, R., Beardsley, B., Lentz, S.: Classical tidal harmonic analysis including error estimates in MATLAB using T-TIDE. Comput. Geosci. **28**(8), 929–937 (2002)
19. Anjyo, K., Lewis, J.P.: RBF interpolation and Gaussian process regression through an RKHS formulation. J. Math. Ind. **3**(2011A–6), 63–71 (2011)
20. Figueiredo, M.A.T.: On Gaussian radial basis function approximations: Interpretation, extensions, and learning strategies. In: Proceedings of 15th International Conference on Pattern Recognition, Vol. 2, pp. 618–621. Barcelona (2000)
21. Hartigan, J.A., Wong, M.A.: Algorithm as 136: a K-means clustering algorithm. J. Roy. Stat. Soc. **28**(1), 100–108 (1979)
22. Simon, D.: Optimal state estimation: Kalman, H∞ and nonlinear approaches. Wiley, Chichester (2006)
23. Gurgel, K.W., Antonischki, G., Essen, H.H., Schlick, T.: Wellen Radar (WERA): a new groundwave HF radar for ocean remote sensing. Coast. Eng. **37**(3–4), 219–234 (1999)
24. Savidge, D., Amft, J., Gargett, A., Archer, M., Conley, D., Voulgaris, G., Wyatt, L., Gurgel, K.W.: Assessment of WERA long-range HF-radar performance from the user's perspective. In: Proceedings of the IEEE/OES/CWTM 10th Working Conference on Current Measurement, Technology, pp. 31–38 (2011)
25. Luettich, R.A., Westerink, J.J., Scheffner, N.W.: ADCIRC: an advanced three-dimensional circulation model for shelves coasts and estuaries, report 1: theory and methodology of ADCIRC-2DDI and ADCIRC-3DL. Technical report, U.S. Army Engineers Waterways Experiment Station, Vicksburg (1992)
26. Blanton, B.O., Werner, F.E., Seim, H.E., Luettich, Jr., R.A., Lynch, D.R., Smith, K.W., Voulgaris, G., Bingham, F.M., Way, F.: Barotropic tides in the South Atlantic Bight. J. Geophys. Res. **109**, C12024 (2004)
27. Szwaykowska, K., Zhang, F.: A lower bound for controlled Lagrangian particle tracking error. In: Proceedings of 49th IEEE Conference on Decision and Control, pp. 4353–4358. Atlanta (2010)

Chapter 5
EasyLoc: Plug-and-Play RSS-Based Localization in Wireless Sensor Networks

Maissa Ben Jamâa, Anis Koubâa, Nouha Baccour, Yasir Kayani, Khaled Al-Shalfan and Mohamed Jmaiel

Abstract Localization based on Received Signal Strength (RSS) is a key method for locating objects in Wireless Sensor Networks (WSNs). However, current RSS-based methods are ineffective at both deployment and operation design levels. First, they usually require a labor-intensive pre-deployment profiling operations to map the RSS to either locations or distances. Second often rely on heavy processing operations. These two design problems limit the possibility of implementing such localization techniques on resource-constrained sensor nodes, and also restrict their scalability and use in practice. In this book chapter, we discuss the challenges and limitations of RSS-based localization mechanisms and we propose, EasyLoc, an autonomous and practical RSS-based localization technique that improves on previous approaches in terms of ease of deployment and ease of implementation, while still providing a reasonable accuracy. EasyLoc is a plug-and-play and fully distributed RSS-based localization method that requires zero pre-deployment configuration. The idea

M. B. Jamâa (✉) · N. Baccour · M. Jmaiel
ReDCAD Research Unit, National School of Engineers of Sfax, Sfax, Tunisia
e-mail: mbenj@redcad.org

N. Baccour
e-mail: nabr@isep.ipp.pt

M. Jmaiel
e-mail: mohamed.jmaiel@enis.rnu.tn

M. B. Jamâa · A. Koubâa · Y. Kayani
COINS Research Group, Prince Sultan University, Riyadh, Saudi Arabia
e-mail: akoubaa@coins-lab.org

Y. Kayani
e-mail: kayaniyasir@rtrackp.com

A. Koubâa · N. Baccour
CISTER Research Unit, Polytechnic Institute of Porto (ISEP/IPP), Porto, Portugal

K. Al-Shalfan
Al-Imam Mohamed bin Saud University, Riyadh, Saudi Arabia
e-mail: kshalfan@imamu.edu.sa

A. Koubâa and A. Khelil (eds.), *Cooperative Robots and Sensor Networks*,
Studies in Computational Intelligence 507, DOI: 10.1007/978-3-642-39301-3_5,

consists in exploiting the available distance information between anchors to derive an online and anchor-specific RSS to distance mapping. We show that, in addition to its simplicity, EasyLoc provides, in the best case, a reasonable average distance error of 2 m in an indoor environment of $30\,m^2$.

Keywords RSS-based localization · Zero pre-deployment configuration · Energy efficiency · Wireless sensor networks

1 Introduction

Localization in Wireless Sensor Networks (WSNs) is a key research area that has been well-studied in the literature. Localization aims at finding the location of an object in its environment. Several techniques have been devised in this respect. For instance, *range-free* techniques [1–3] estimate the location of a sensor node without calculating the distance to anchors but rather relying on other logical information including radio connectivity, anchor proximity and sensing capabilities. On the other hand, *range-based* techniques [4, 5] rely on the estimation of distances between the unknown node and anchors to infer the position of the unknown node using lateration techniques. These techniques are known to achieve better localization accuracy than range-free solutions. However, they have the limitation of increasing system complexity in terms of ranging hardware (e.g. additional hardware usage such as ultrasound emitters, directional antenna), careful calibration and environment profiling.

Distance estimation is typically achieved through three main approaches: (i) *time-based approach*, which consists in calculating the distance by measuring the radio signal propagation time between two nodes, such as Time-of-Arrival (TOA), Round-trip Time of Flight (RTOF), Time Difference of Arrival (TDOA), (ii) *angle-based approach*, which relies on computing the angle (or the direction) of the line connecting an unknown node to an anchor with respect to some reference direction, such as Angle-of-Arrival (AOA), and (iii) *RSS-based approach*, which consists in finding a mapping between the received signal strength and the distance to a node or its location.

Although time-based and angle-based approaches are capable of achieving high accuracy level, several factors hinder their deployment expansion in resource-constrained low-power networks, such as WSNs. Thus, RSS-based methods are the most appealing method for locating objects in sensor network with a minimal cost. Indeed, RSS-based localization represents a non expensive solution. First, it does not require any additional ranging hardware since it relies on the built-in wireless transceivers. Second, it exhibits a low computational complexity as RSS can be directly read from the wireless transceiver (e.g. CC2420 transceiver of TelosB/MICAz motes). Furthermore, unlike time-based approaches, RSS-based localization does not need any synchronization services. However, classical RSS-based localization techniques typically induce a labor-intensive pre-deployment phase as the RSS behavior is heavily dependent on the environment [6]. Such pre-deployment phase represents a major handicap constraining the large scale deployment of these techniques in real-world applications.

In this chapter, we tackle the aforementioned problems and come up with a new RSS-based localization technique, coined as EasyLoc, that is easy to deploy in any (indoor/outdoor) environment, while still satisfying performance requirements in terms of scalability, low computation complexity, robustness against noise, energy-efficiency and accuracy.

The remainder of this chapter is organized as follows. In Sect. 2, we provide a background on the existing RSS-based localization techniques. In Sect. 3, we discuss their limitations. Section 4 introduces EasyLoc, our proposed RSS-based localization method. In Sect. 5, we present an extensive experimental study to evaluate the performance of EasyLoc and draw the main learned lessons. Finally, Sect. 6 concludes the chapter and presents insights into our future work.

2 RSS-Based Localization

RSS-based localization consists in creating a mapping between the distance or location and the Received Signal Strength (RSS). RSS-based localization is built upon the assumption that the signal attenuates when the distance increases. Theoretically, the relation between the RSS and the distance is captured by the well-known path loss log normal shadowing model expressed as:

$$RSS(d)[dB] = RSS(d_0) - 10\eta \log(d/d_0) + X_{noise}[dB] \quad (1)$$

where $RSS(d)$ is the Received Signal Strength at distance d from the sender, $RSS(d_0)$ is the Received Signal Strength at a reference distance d_0 fixed and known in advance, η is the path loss exponent that measures the rate at which the RSS decreases with respect to the distance, and X_{noise} is a zero-mean Gaussian random variable with a variance σ^2, which is referred to as the shadowing variance. Both η and σ^2 are environment-dependent.

Exploiting RSS measurements for localization has been always an attractive approach for WSN designers because RSS is easily accessible from all available sensor motes built-in wireless transceivers. However, this does not come without cost as RSS-based localization is known to be inaccurate due to the high RSS variability caused by random multi-path effects. These effects result from the obstruction of physical objects during the signal propagation making the signal weaker and likely to be distorted. An empirical analysis of the impact of RSS in localization in WSNs has been reported in [7] where the authors have drawn the following three observations: (i) The distribution of the RSS is not necessarily Gaussian and it depends on the environment and on the transmission power, (ii) The RSS variability is typically (very) high in particular when the propagation is disturbed by the multi-path effects and interferences and (iii) The non isotropic behavior of signal propagation is the cause of the spatial RSS variability.

In the literature, several RSS-based localization techniques were proposed. These techniques can broadly be classified into two categories: *map-based* and *model-based*. Map-based techniques such as [8–10] rely on fingerprinting the environment through extensive pre-deployment measurements. In general, these mechanisms are composed of two phases: *the training phase* and *the positioning phase*. The training phase consists in subdividing the deployment area into cells, measuring the RSS at each cell and then building a map of radio fingerprints (also referred as signatures) where each fingerprint refers to a unique cell in the deployment area. In the positioning phase, an unknown node localize itself by computing the cell that best fits the collected RSS measurements. Maximum-likehood [8] or K-Nearest Neighbors [9] can be used to find the best fit. On the other side, model-based techniques such as [11–14] aim at establishing a mathematical model capturing the variation of the RSS as a function of the distance. The most three common models are (i) *Free space propagation model* that supposes the absence of multi-path effects and that communicating node are in line of sight, (ii) *Two-Ray ground model* that improves the first model by considering reflection effects and (iii) *Log-Distance Path Loss model* that takes into account stronger attenuation by defining a path loss exponent. This model is usually suffixed by a Gaussian random variable in order to take into account obstruction effects (Log-Normal Shadowing Model). An exhaustive taxonomy of model-based techniques can be found in [15]. For both map-based and model-based mechanisms, it is central to pass through an offline and tedious environment profiling phase to collect empirical data to map the distance (or cells) to RSS. This process is usually cumbersome, and at the end, the resulting static mapping is prone to error as it is not robust to the dynamics of the environment (moving persons, environmental conditions, etc.). Adapting the mapping to each change is costly, in particular for map-based techniques, as it requires the iteration of the profiling operation. To cope with the tedious and cumbersome environment profiling phase, recent RSS-based localization studies, such as [15–23], have provided dynamic RSS to distance/location mapping solutions which usually rely on a runtime calibration process. In the following section, we discuss the most relevant related works.

3 Problem Statement

Although the literature on RSS-based localization is mature, the majority of existing solutions are not easy to deploy, as they require manual training phase. Also, some solutions are energy-greedy and are not appropriate to resource constrained sensor networks. For instance, works in [16, 17, 19, 20], have considered high-power radios (e.g. WiFi, GSM), which are fundamentally different from low-power radios used in WSNs [6]. Furthermore, works in [15–18, 20, 21, 23] rely on centralized approaches as they require a global knowledge about the network and the environment characteristics. Typically, a central server with high processing capabilities is needed to perform intensive and heavy-loaded processing operations on empirical data such as iterative Matrix manipulations (e.g. SVD, LMS) [17, 18, 20, 21], probabilistic metaheuristic [15] and genetic algorithms [20].

For instance, authors in [20] have proposed an indoor localization algorithm for WiFi-based networks. The algorithm is centralized and computations are done at a server level. To localize itself, an unknown node collects RSS measurements from APs and reports them to the server. This latter uses collected data to simultaneously learn the parameters of the radio propagation model (Log Distance Path Loss model (LDPL)), determine APs locations and localize unknown nodes. LDPL parameters and APs locations are derived by solving a linear system using both genetic and gradient descent algorithms. To reduce the search space, some constraints are assumed.

Authors in [16] have proposed an online calibration process of RSS to distance mapping for a WiFi-based network. The calibration process is done at a central server. This latter relies on the use of inter-anchors (APs) exchanged traffic in order to generate for each anchor multiple RSS to distance mapping functions. This method is memory greedy and shares the drawbacks of centralized approaches. Indeed, although centralized approaches are known to provide *(near)* optimal solutions, they are not scalable and also restrict the autonomous deployment. In addition, computations are complex and memory greedy which inhibit the possibility of their implementation in resources constrained sensor nodes. Furthermore, extra traffic needed to report RSS measurements to the central server has a negative impact on sensor node internal energy.

Although several works, such as [16–21] have reduced the deployment phase complexity, they still suffer from complex computations and are generally restricted by a centralized architecture. In particular, authors in [22] have considered practical issues in their design, but their approach was simply validated using MATLAB simulations and offline empirical data analysis. This raises questions about the possibility to implement the approach on resource-constrained sensor node.

In our work, we consider the ease of deployment and simple and distributed computations as main concerns. We thus, devise a solution with low complexity that we effectively implement on TelosB motes. To meet this objective, we propose EasyLoc, a lightweight, practical, scalable and distributed RSS-based localization method. Our problem can be formulated as follow: *Given a set of existing anchors with known location, how to design a RSS based localization method for low power sensor networks that (1) is sufficiently accurate (2) can be easily deployed with no fingerprinting or pre-deployment profiling (3) adaptive to environmental changes, (4) is fully distributed and (5) does not overwhelm the sensor node stringent resources.* We present our solution in the next section.

4 EasyLoc

EasyLoc is a model-based plug-and-play RSS-based localization method that has several design requirements. First, it mitigates the effect of hardware imperfections by establishing an anchor-specific RSS to distance mapping. Moreover, it is easy to deploy as it has not any pre-deployment calibration phase and builds its mapping in runtime. This mapping is dynamically updated, thus allowing for better localization accuracy. Finally, it is easy to implement on resources constrained devices. As depicted in Fig. 1, EasyLoc comprises two main phases: *Online calibration phase* and *Localization phase*.

4.1 Online Calibration Phase

The objective of the online calibration phase is to allow each anchor to build its own RSS to distance mapping by exploiting the knowledge of the distance to each of its neighbor anchors. This eliminates the need for building a global RSS to distance mapping through cumbersome environment profiling. *Algorithm 1* in Table 1 presents the RSS to distance mapping process running in each anchor. Each anchor collects a raw of RSS measurements via exchanging a number of probe messages (*prbMsg*) with its neighbor anchors (Step 1 in Fig. 1a). Then, it filters them in order to reduce the RSS time-variability and stores the filtered value (*ARSS*) in its neighbor table (*neighTB*) (Step 2 in Fig. 1a). In the last step, the anchor determines a model (Eq. 1) mapping *RSS* and distance by fitting stored data. For sake of low computational complexity, EasyLoc uses Least-Squares linear regression (Step 3 in Fig. 1a) as a fitting function and a simple averaging filter to estimate the noise-free RSS value (*ARSS*). Nevertheless, we have tested other (more complex) functions for regression and filtering, namely, Random Sample Consensus (RANSAC) [24], and polynomial regression respectively

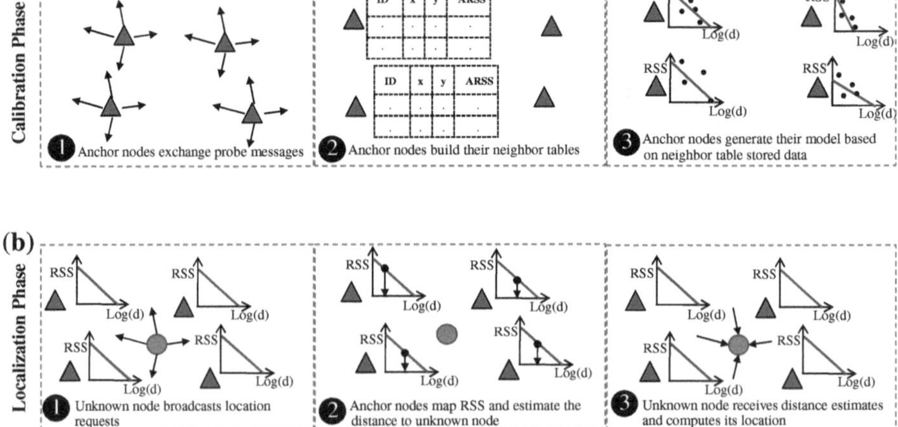

Fig. 1 EasyLoc calibration and localization phases. **a** Online calibration phase. **b** Localization phase

The RSS to distance mapping line coefficients a and b are extracted by minimizing the sum of squared residuals q, expressed in the following equation

$$q = \sum_{i=1}^{n} \left[RSS_i - (a + b \times \log(d_i))\right]^2 \tag{2}$$

where d_i and RSS_i are respectively the distance to the ith neighbor anchor and its filtered RSS measurement. a and b are solved by the following equations (Eqs. 3 and 4).

$$\frac{\partial q}{\partial a} = -2\sum_{i=1}^{n} \left[RSS_i - (a + b \times \log(d_i))\right] = 0 \tag{3}$$

$$\frac{\partial q}{\partial b} = -2\sum_{i=1}^{n} \log(d_i)\left[RSS_i - (a + b \times \log(d_i))\right] = 0 \tag{4}$$

RSS to distance model is periodically updated each time a raw of new couples (RSS,distance) feeds an anchor. This makes EasyLoc adaptive to environment changes as well as radio changes.

4.2 Localization Phase

This phase aims at determining the absolute location of an unknown node using a lateration technique, such as Min-Max [25] or Weighted Centroid (WC) [26].

Algorithm 2 Table 1 represents the localization process running at the unknown node and *Algorithm 3* Table 1 describes the localization process running at each anchor. Figure 1b illustrates the localization steps. First, unknown node triggers the localization process by broadcasting a burst of w location requests (*locReq*) to neighbor anchors (Step 1 in Fig. 1b). Each anchor, on its side, measures signals strengths of received *locReq*, averages them (*RSSAvg*) and uses its mapping model to estimate the distance to the unknown node (Step 2 in Fig. 1b). After that, a location response packet is sent to the unknown node where the anchor has included the estimated distance and its location coordinates (Step 3 in Fig. 1b). Subsequently, unknown node waits for at least three location response packets from distinct anchors in order to determine its location. In the first EasyLoc instance, Min-Max algorithm [25] was used as a lateration technique.

Table 1 EasyLoc algorithms

Algorithm 1. Online Calibration Phase in an Anchor Node

Input: Timer1 $t1$, probe burst size *n*, averaging window *Aw*
Output: Anchor-specific RSS to distance linear mapping

```
thread probeTransmission () {
  if (t1 seconds have elapsed) then
    broadcast a burst of n prbMsg containing
    the ID and the location coordinates X and Y
  end if
}
thread probeReception () {
  if (a prbMsg is received) then
    update neighbor table (neighTB) based on received data
    (ID,X,Y, RSS[], AvgRSS, distance, Nb of received prbMsg)
  end if
}
thread mappingUpdate () {
  if (Nb of received prbMsg ≥ (Aw)) then
    Filter RSS[] vector values, Update AvgRSS and
    Compute the parameters of the RSS to distance mapping model
    // model: RSS = a + b × log10(distance)
    Clean RSS[] vector and Reset number of received prbMsg
  end if
}
```

Algorithm 2. Unknown Node Localization Process

Input: Location request burst w, Inter-Packet Arrival time IPI
Output: Location of the unknown node

```
send successively w location requests (locReq) with a period IPI
// locReq message contains Unknown node ID and IPI information
wait()
if (at least three distance estimates ≤ Dist_th are received) then
  estimate location using lateration technique (e.g. Min-Max).
end if
```

(continued)

We point out that unknown node must eliminate outliers (if any), which are distances greater than a certain environment-specific threshold $Dist_{th}$, and must rely only on relevant distances. This helps to filter out wrong overestimated distances that might compromise the localization accuracy. The maximum distance threshold is environment-dependent and can be set to the maximum distance between any two nodes in the environment.

Table 1 (continued)

Algorithm 3. Anchor Node Localization Process
Input: Timer1, Timer2 $t2$, RSS-to distance mapping
Output: Distance to the unknown node

if (a *locReq* is received) **then**
  stop Timer1 and Timer2
  set Timer2 period to $t2$ ($t2 = 2 * IPI + \epsilon$)
 // ϵ varies from an anchor to another
  increment Nb of received *locReq*
end if
if ($t2$ seconds have elapsed without receiving a *locReq*) **then**
  compute the average RSS (*RSSAvg*) of received *locReq*
  determine the distance corresponding to *RSSAvg* using RSS-to-distance mapping
  send the measured distance to unknown node
  stop Timer2 and relaunch Timer1
end if

5 Experimental Evaluation Performance

5.1 Experiments Design

We have implemented EasyLoc on TelosB motes using TinyOS to evaluate its performance and to demonstrate its practical-aspect and effectiveness through real experiments. Experimental data collection and data analysis of EasyLoc experiments were carried out using *iLoc* tool [27], a tool that we designed for automating localization experiments with WSNs. *iLoc* comprises two independent applications: (i) *iLoc-Controller*, which is a Java application responsible for the experimental data collection and experiments configuration, and (ii) *iLocAnalyzer*, which is a MATLAB application that allows empirical data analysis to assess the statistical properties of RSS-based localization techniques. Our experiments were conducted in two indoor **interference-free** and **static** environments with different extents: *Small* environment of $2\,\text{m}^2$ area size ($1\,\text{m} \times 2\,\text{m}$) and *Large* environment of $30\,\text{m}^2$ area size ($4.4\,\text{m} \times 6.8\,\text{m}$), where a set of anchors were deployed. The objective of considering two different-size environments is to assess the impact of area size on localization error. In each scenario, 12 unknown nodes locations were estimated using the Min-Max [25] algorithm. For each unknown node location, a burst of w location requests (*locReq*) are transmitted to anchors. To ensure the reliability of localization error analysis, this operation is repeated for *NbBurst* times. Figure 2 depicts how sensor nodes are deployed and Table 2 illustrates the experimental settings.

The performance of EasyLoc was evaluated by analyzing the (i) *distance error*, which is the difference between the real and the estimated distances between anchors and the unknown node, (ii) *location error*, which is the difference between the real

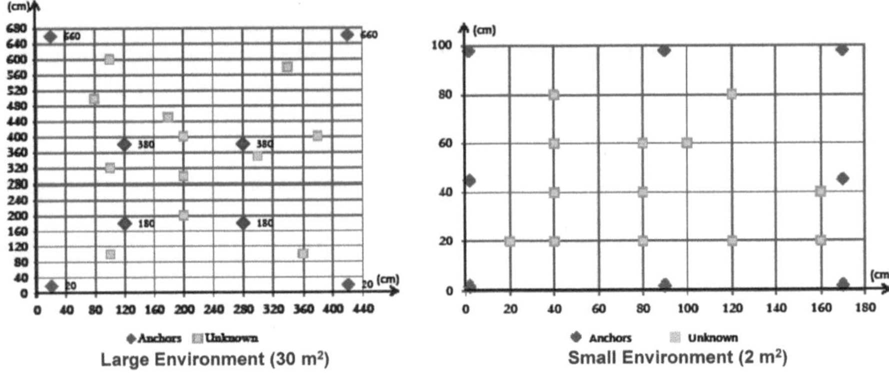

Fig. 2 Sensor nodes disposition in deployment area. **a** Large environment ($30\,m^2$). **b** Small environment ($2\,m^2$)

Table 2 Experimental settings

Environment extents	Impact of	NbBurst	w	NbA	Txp	Tx channel
Small	Txp	5	10	8	0, −15, −25	26
	NbA			4, 6, 8	0	
Large	Txp	10	5	8	0, −15	
	NbA			4, 8	0	

and estimated location of the unknown node. Errors were analyzed through their Cumulative Distribution Functions (CDF), their averages, Standard Deviations (StD) and the percentage of outliers, which is the percentage of discarded distances due to exceeding the maximum threshold. The impacts of both transmission power (*Tx power*) and number of anchors (*Nb Anchors*) on distance and location errors were investigated.

5.2 Experimental Results

The experimental study reveals several results that we summarize in the following observations.

Observation 1. Increasing the transmission power reduces the location/distance errors. However, it also increases the percentage of outliers and leads to the greatest maximal error. It is clear from Figs. 3 and 6 that the accuracy of both estimated location and distance is worse for lower transmission powers, and for smaller number of anchors. Indeed, it is also interesting to observe that the Cumulative Distribution Function (CDF) curves of location and distance errors grow more smoothly with low transmission powers (i.e. −15 and −25 dBm) than with the maximum transmission power (0 dBm). Furthermore, it can be noted from Figs. 4 and 7 that the average of

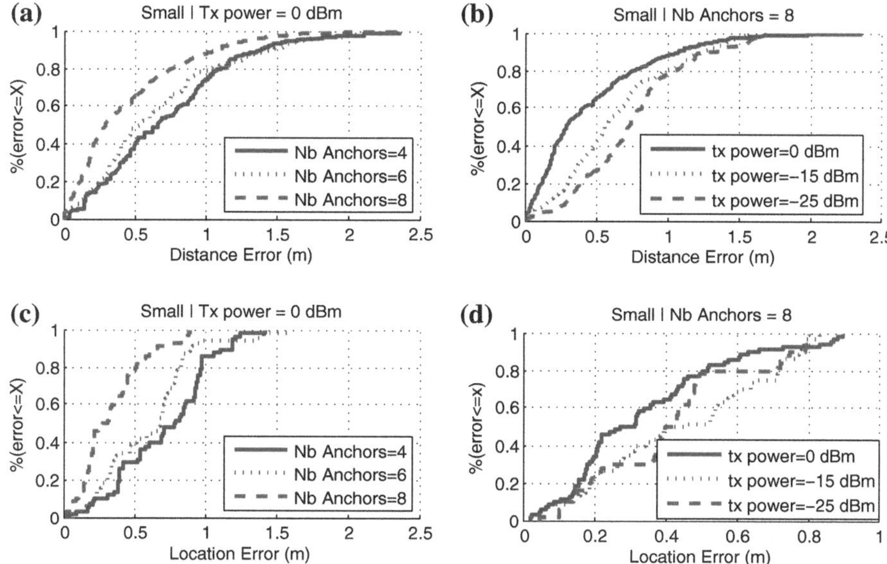

Fig. 3 Small environment: localization accuracy. **a** Impact of number of anchors on distance errors. **b** Impact of transmission power on distance errors. **c** Impact of number of anchors on location errors. **d** Impact of transmission power on location errors

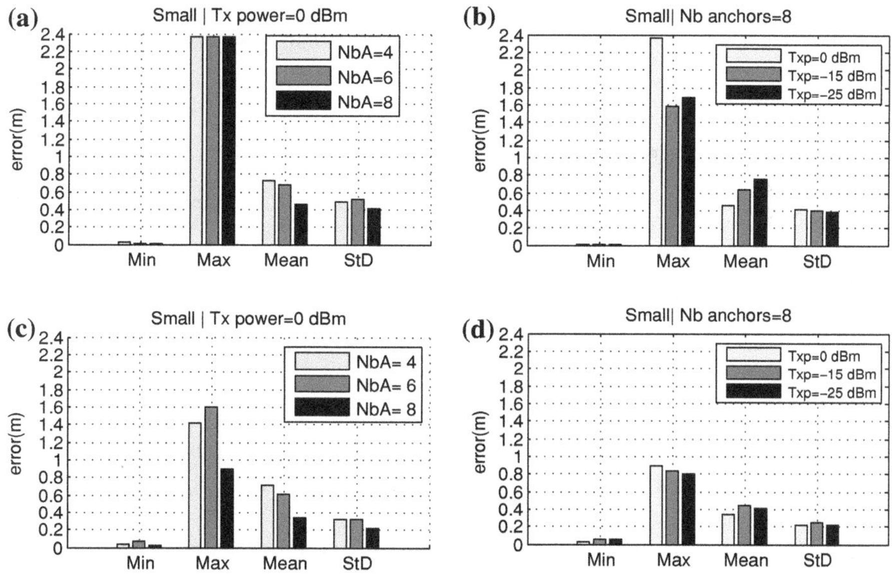

Fig. 4 Small environment: localization errors: Min, Max, Mean, StD. **a** Impact of number of anchors on distance errors. **b** Impact of transmission power on distance errors. **c** Impact of number of anchors on location errors. **d** Impact of transmission power on location errors

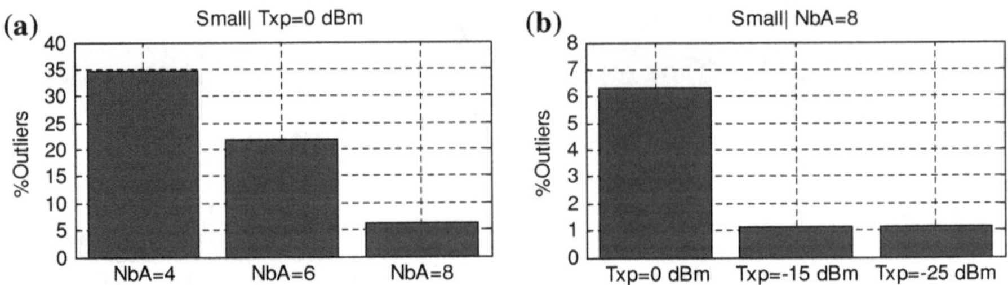

Fig. 5 Small environment: percentage of outliers. **a** Impact of number of anchors on distance errors. **b** Impact of transmission power on distance errors

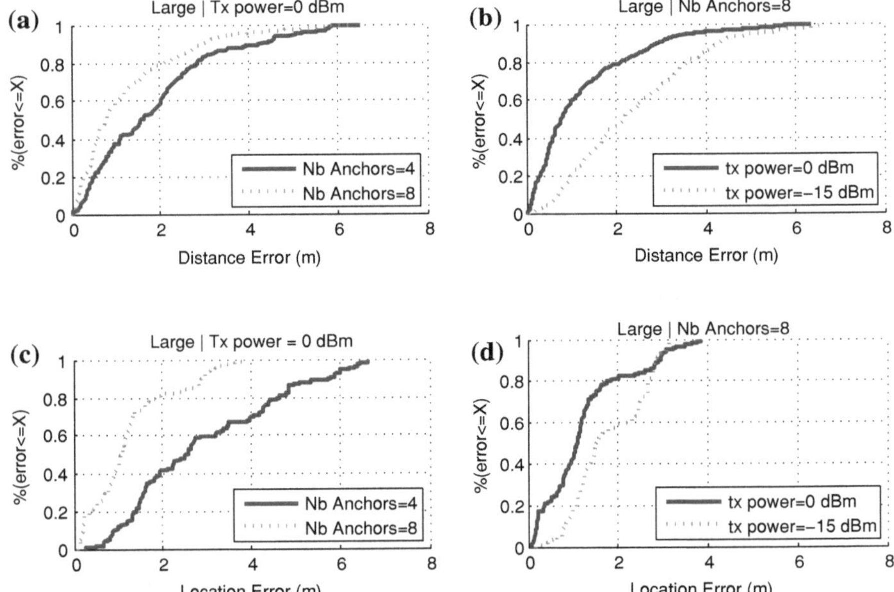

Fig. 6 Large environment: localization accuracy. **a** Impact of number of anchors. **b** Impact of transmission power. **c** Impact of number of anchors. **d** Impact of transmission power

location and distance errors are lower with 0 dBm transmission power. This observation illustrates the negative impact of low-power links on achieving good localization accuracy, which makes localization more challenging in WSNs. However, increasing the transmission power for better accuracy would not be the convenient solution, as it also increases the probability of having more oultiers as it is depicted in Figs. 5 and 8.

Observation 2. EasyLoc provides an acceptable distance/location accuracy especially when the number of anchors and the transmission power are high. It can be observed from Figs. 4d and 6d that the average of location error varies between 33 and 71 cm in the small environment, and between 1.25 and 2.9 m in the large environment. We also observe from Figs. 4d and 6d that, in best case (i.e. Tx power = 0

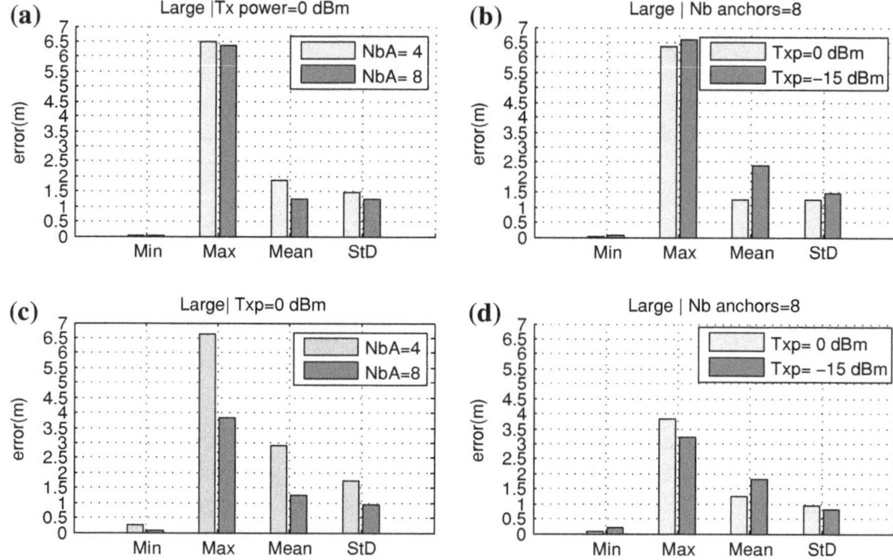

Fig. 7 Large environment: localization errors: Min, Max, Mean, StD. **a** Impact of number of anchors on distance errors. **b** Impact of transmission power on distance errors. **c** Impact of number of anchors on location errors. **d** Impact of transmission power on location errors

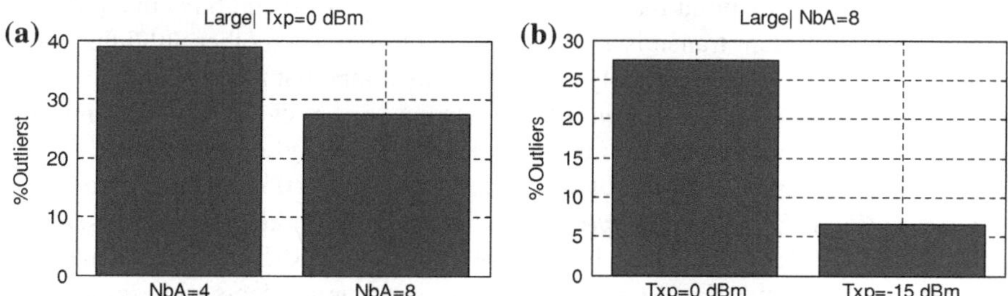

Fig. 8 Large environment: percentage of outliers. **a** Impact of number of anchors on distance errors. **b** Impact of transmission power on distance errors

dBm and Nb anchors = 8), 80 % of location errors are less than 0.5 m in the small environment, and 80 % of location errors are less than 2 m in the large environment. On the other hand, in the worst case (i.e. Nb anchors = 4), Figs. 4c and 6c show that 80 % of location errors are less than 1 m in the small environment, and 80 % of location errors are less than 3 m in the large environment. This accuracy decrease can be justified to larger errors in distance estimates in comparison to the best case. For instance and as illustrated in Fig. 4a, 80 % of distance errors, in small environment and in the best case, are less than 70 cm whereas in the worst case, only 55 % are less than 70 cm. The same observation is also held for large environment as shown in Fig. 6a. Distance estimates accuracy depends on the rigor of the RSS to distance

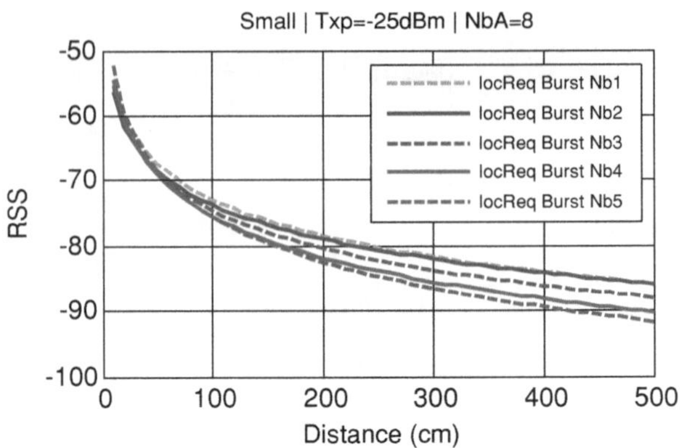

Fig. 9 Mapping curves evolution for one anchor

mapping model derived by anchors. Indeed, higher is the number of anchors, more the regression method, used to derive the RSS to distance mapping model, is fed by additional set of (RSS, distance) couples and thus better the regression behaves. It turns out that EasyLoc, in addition to its ease of deployment, can achieve a reasonable localization accuracy when both number of anchors and the transmission power are high.

Observation 3. When the number of anchors is high, Min-Max lateration technique reduces the impact of distance errors on location errors, regardless the transmission power. The interesting point that can be drawn from Figs. 4d and 6d is that the location errors for different transmission powers are close to each other although corresponding distance errors are clearly different. This infers that increasing the number of anchors will be able to compensate the distance errors even if the transmission power is low. This represents an interesting feature for EasyLoc as it allows the use of low-power transmission without significantly compromising location errors.

Observation 4. EasyLoc is adaptive to radio changes. It is known that RSS significantly varies over long time spans [6]. It implies that static RSS-to-distance mapping will change over time. EasyLoc addresses this issue as it adapts to radio changes via periodic calibration refresh (Algorithm 1 line 13). Figure 9 shows how an anchor mapping curve evolves over time in response to radio changes. Each of these curves is used to respond to a burst of *locReq* messages. These bursts are sent consecutively in time.

5.3 EasyLoc: Error Analysis and Potential Improvements

In what follows, we analyze the major factors impacting EasyLoc performance and we provide hints for performance improvements.

Radio Antenna Most of commercial off-the-shelf sensor motes embed an internal antenna with poor gain which makes the radiation pattern significantly anisotropic

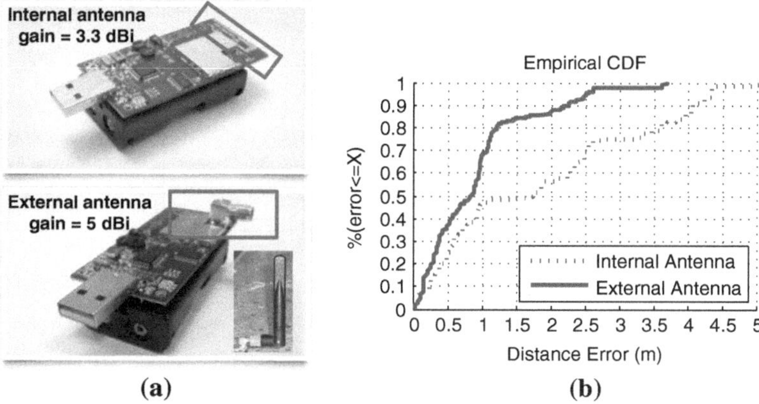

Fig. 10 Impact of the use of antenna with good gain. **a** Original and cloned TelosB motes. **b** Experiment results: distance error

(i.e. antenna does not radiate equally in all directions). The use of an external antenna with higher gain improves sensor motes sensitivities and leads to a (*near*) uniform radiation pattern. Importantly, adding an external antenna has neither impact on sensor motes energy since antenna are passive nor on their cost since antenna are cheap. To assess the impact of the use of such antenna on localization accuracy, we have performed two experiments in (8 m × 2 m) indoor environment (a corridor). In the first experiment, we have used original TelosB motes that integrate 3.3 dBi antenna, while in the second, we have used a cloned TelosB motes with 5 dBi external antenna fixed on the SMA connector. In each experiment, we considered a network of 9 anchors and 5 unknown nodes placed at different locations. Each unknown node triggers the localization process by sending a burst of 5 location requests and then collects distance estimates from neighbor anchors. This operation is repeated 10 times in order to improve the reliability of results.

As illustrated in Fig. 10, we observe that about 85 % of distance errors are less than 1.5 m when 5 dBi antenna are used against only 50 % with default 3.3 dBi antenna. This demonstrates that the negative impact of RSS spatial variability on localization accuracy can be alleviated by using enhanced gain antenna.

RSS measurements filtering technique Received Signal Strength (RSS) measured between two nodes at fixed distance is known to be very variable over time. Filtering RSS is a common solution to smooth this time variability. Currently, EasyLoc filters RSS measurements using a simple averaging method. Nonetheless, more sophisticated filters can also be adopted namely (1) Kalman filter (KF) which is a well-known filtering method, that has been widely used in the area of autonomous or assisted navigation, (2) weighted moving average, which assigns weights to both old and recently computed RSS average (3) the Savitzky-Golay[28] filter which performs a polynomial regression on a small set of consecutive data points and then sets the

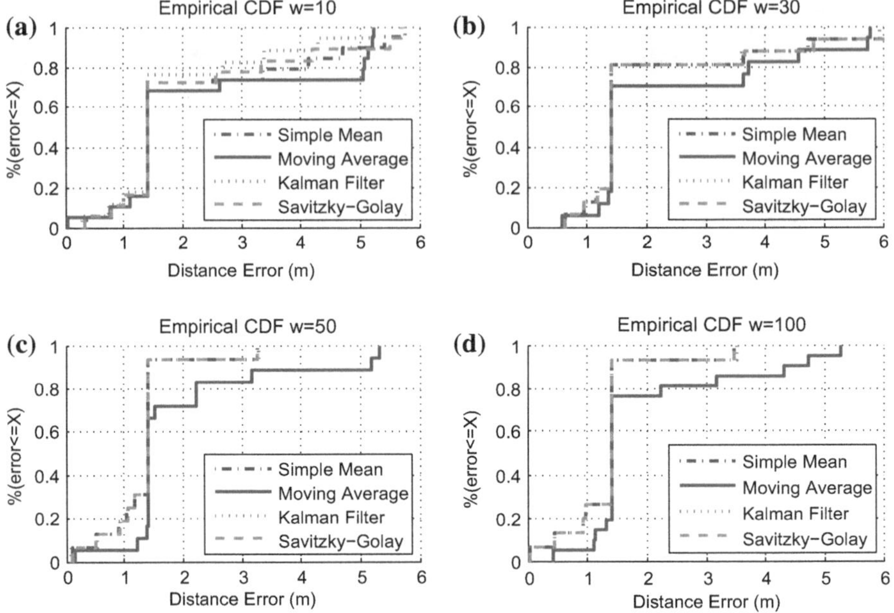

Fig. 11 Cumulative distribution function of the distance error for different filtering techniques using Least-Squares regression. **a** Filtering window size = 10. **b** Filtering window size = 30. **c** Filtering window size = 50. **d** Filtering window size = 100

central point of the generated curve as the filtered value. To find out which of these filtering techniques is more resilient to RSS time variability and thus leads to better localization accuracy, we conducted the following experiment: a set of 9 anchors and 5 unknown nodes deployed on the ground in 2 m × 8 m indoor environment. In this experiment, we collected all RSS measurements gathered by a **particular anchor** upon the reception of probe messages (*prbMsg*) or location requests (*locReq*). Then using MATLAB, we replicated the behavior of this anchor as if it was executing EasyLoc. Basically, we recomputed the RSS/distance model based on filtered data and we re-estimated distances to unknown nodes from which this anchor has heard *locReq* messages. In our study, we also evaluated the impact of filtering window size (w) on the filter performance. Figure 11 shows the Cumulative Distribution Function (CDF) of estimated distances error for the four filtering techniques using four different filtering window sizes and Fig. 12 shows the minimum, the maximum, the mean and the standard deviation of the estimation error.

From Fig. 11, we notice shows that more the filtering window size is large more the localization error is reduced. This demonstrates that transient RSS variation can be canceled by the filter, only if a big dataset is offered. Furthermore, we notice that studied filtering techniques have roughly similar performance, expect for weighted moving average that shows the worse localization accuracy. Indeed, weighted moving average filter has the highest estimation error mean and it is less stable since it has

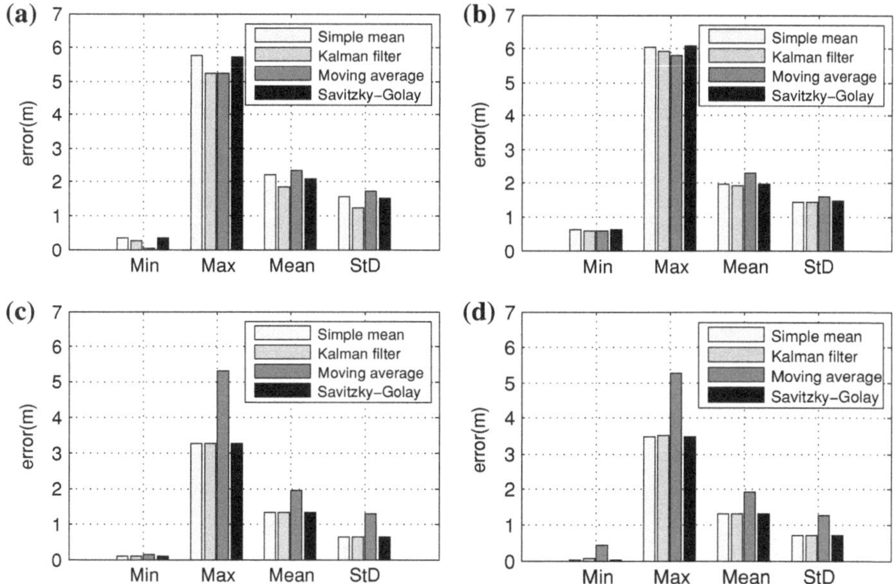

Fig. 12 Localization errors: Min, Max, Mean, StD for different filtering techniques using Least-Squares regression. **a** Filtering window size = 10. **b** Filtering window size = 30. **c** Filtering window size = 50. **d** Filtering window size = 100

the highest estimation error variation (StD). Another interesting observation can be drawn from Fig. 11: the localization error for the four filtering technique is the same for both filtering widow sizes (w) 50 and 100. This observation shows that studied filters have reached the stationary state. Such observation is very useful in order to (i) minimize the number of RSS measurements to be stored by the anchor and (ii) to maintain acceptable localization accuracy.

RSS and distance mapping In EasyLoc, the accuracy of estimated distances dependents on how well collected RSS measurements match the RSS to distance mapping model of anchors. For sake of simplicity and ease of implementation, EasyLoc, currently computes this mapping model using linear Least-Squares regression. Nonetheless, other methods such as the Random Sample Consensus (RANSAC) [24] can also be used. To analyze how the selected regression method affects EasyLoc performance, we experimentally compared the localization errors gathered by RANSAC and Least-Squares. Both methods map the RSS linearly to distance. The experiment setup is the same as the one described in Sect. 5.3.

From Figs. 11–15, we clearly observe that RANSAC leads to better localization accuracy than Least-Squares. Indeed, RANSAC has a negligible percentage (about 3 %) of outliers and about 60 % of localization error are less than 1 m. In contrast, Least-Squares often fails to corrected map RSS to distance since about 60 % of estimated distances are discarded and about 80 % of localization error are greater

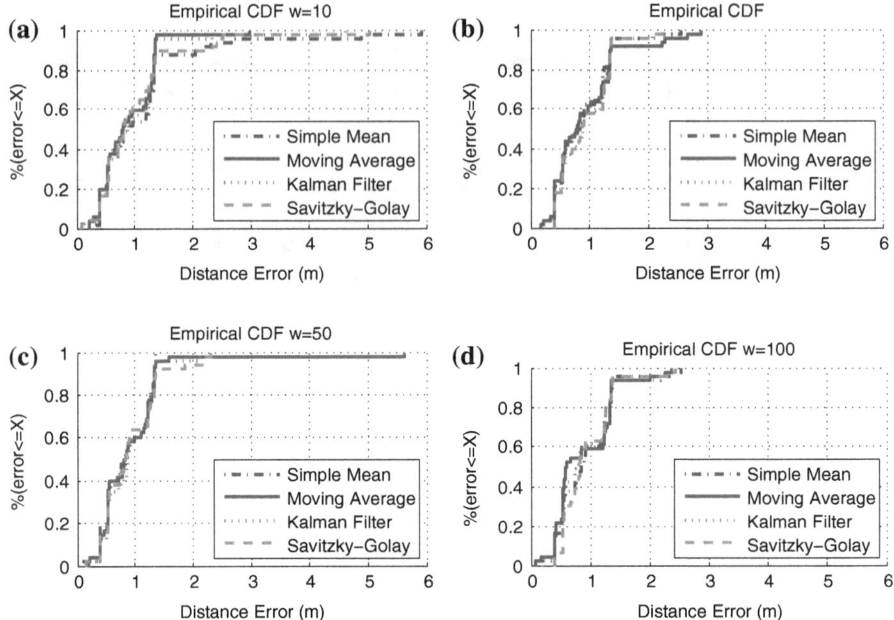

Fig. 13 Cumulative distribution function of the distance error for different filtering techniques using RANSAC regression. **a** Filtering window size = 10. **b** Filtering window size = 30. **c** Filtering window size = 50. **d** Filtering window size = 100

than 1 m. We justify this result by the weakness of Least-Squares in distinguishing which of collected sample is good or bad as it assumes that the noise distribution is the same for the all samples.

Anchors placement layout It is well known in the literature that anchor placement layout has a significant impact on the localization accuracy [29, 30]. Here, we study this impact on the effectiveness of EasyLoc trilateration method. Currently, Min-Max algorithm is used in order to derive unknown nodes locations. Min-Max is a range-free algorithm that gives a coarse grained localization accuracy. It estimates the location of unknown node by (1) computing all squares centered at anchors and with a vertex length equal to their distance to the unknown node, (2) computing the bounding box resulting from the intersection of all squares and (3) computing the center of the bounding box. The edges of the bounding box are computed as follows:

$$[\max(x_i - d_i), \max(y_i - d_i)] \times [\min(x_i + d_i), \min(y_i + d_i)] \tag{5}$$

where $[x_i; y_i]$ are the location coordinates of the anchor i and d_i represents the distance separating it to the unknown node.
In our study, we have analytically re-estimated the location of 12 unknown nodes by using correct distance to anchors measurements for two different anchors layouts.

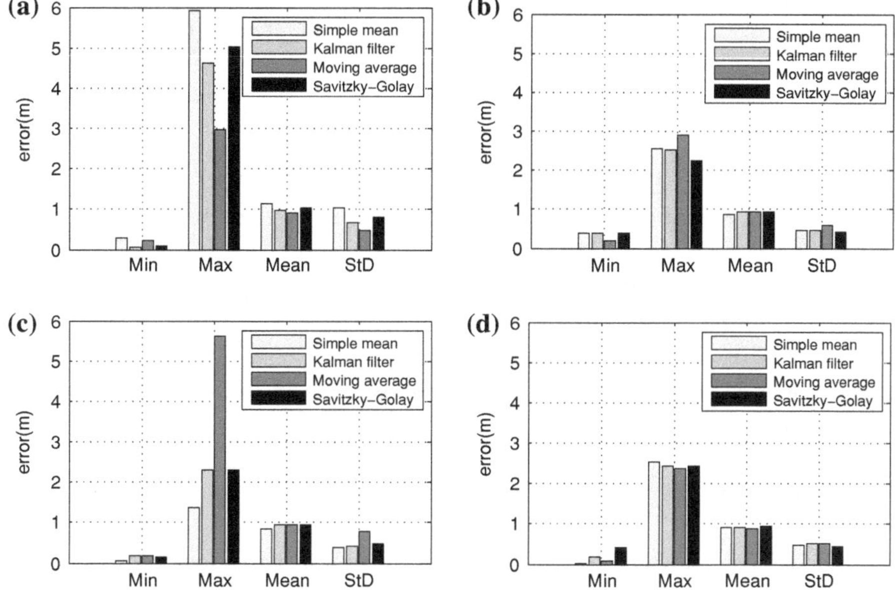

Fig. 14 Localization errors: Min, Max, Mean, StD for different filtering techniques using RANSAC regression. **a** Filtering window size = 10. **b** Filtering window size = 30. **c** Filtering window size = 50. **d** Filtering window size = 100

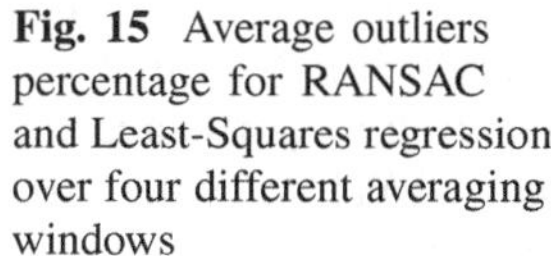
Fig. 15 Average outliers percentage for RANSAC and Least-Squares regression over four different averaging windows

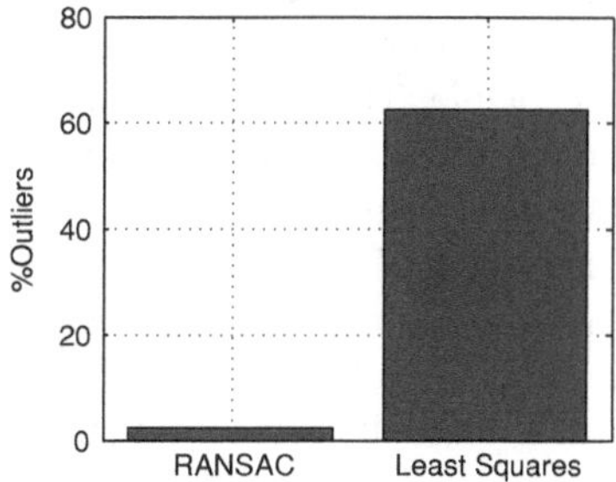

Figure 16 illustrates these layouts together with the resulting localization accuracy. It can be seen from Fig. 16(b) that Min-Max algorithm performance may significantly decrease when anchors are not properly placed in the network. Indeed, although correct distance to anchors measurements are used by Min-Max, localization error is not null and may reach 2 m in a 7 m $\times$ 4.5 m area. Thus, in order to enhance EasyLoc performance, we need to find the best anchor layout that leads to an acceptable localization accuracy. The problem becomes more difficult since the placement layout of listenable anchors highly depends on network connectivity.

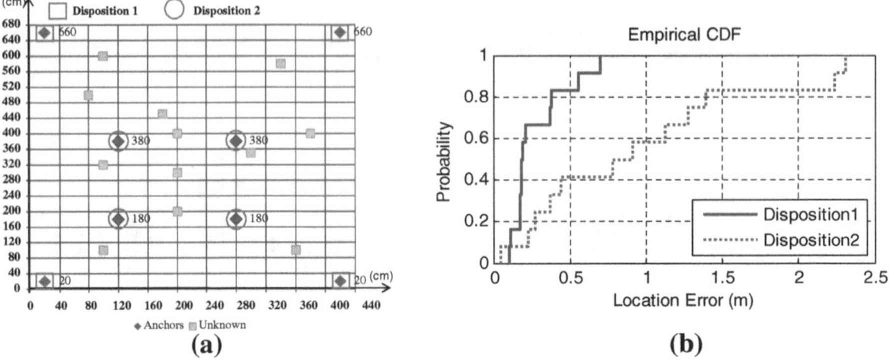

Fig. 16 Impact of anchor placement layout. **a** Anchor placement layouts. **b** Localization accuracy

6 Conclusion

In this book chapter, we proposed EasyLoc, a novel RSS-based localization method that requires zero pre-deployment profiling operations. The main idea was to exploit the knowledge of inter-anchors distances to derive an anchor-specific RSS to distance mapping. This demonstrates that RSS-based localization can be easily deployed, in contrast to traditional cumbersome methods. The performance of EasyLoc was experimentally evaluated through extensive measurements in two indoor environments. Based on our experimental study, we found that EasyLoc accuracy is not sufficiently acceptable. This motivated us to grasp the main factors impacting its performance. To meet this objective, we conducted an analytical study using MATLAB where we assessed the impact of anchor placement layout on EasyLoc performance as well as the impact of gain of antenna, filtering method of RSS measurements and the way to map RSS to distance. In our future work, we plan to improve the current EasyLoc instance by addressing the impact of aforementioned factors and then to assess its performance by considering large-scale and more realistic deployment environments.

Acknowledgments This work is funded by the R-Track project [31] under the grant 8-INF-2008 of the National Plan for Sciences and Technology (NPST), managed by the Science and Technology Unit of Al-Imam Mohamed bin Saud University and by King AbdulAziz Center for Science and Technology (KACST).

References

1. He, T., Huang, C., Blum, B.M., Stankovic, J., Abdelzaher, T.: Range-free localization schemes for large scale sensor networks. In: MobiCom '03: Proceedings of the 9th Annual International Conference on Mobile Computing and Networking, San Diego (2003)
2. Stoleru, R., He, T., Stankovic, J.A.: Range-free localization. In: Chapter in Secure Localization and Time Synchronization for Wireless Sensor and Ad Hoc Networks, vol. 30 (2007)

3. Zhong, Z.: Range-free localization and tracking in wireless sensor networks. Phd dissertation, University of Minnesota. http://www-users.cs.umn.edu/zhong/papers/dissertation.pdf (2010)
4. Amundson, I., Koutsoukos, X.D.: A survey on localization for mobile wireless sensor networks. In: Proceedings of the 2nd International Conference on Mobile Entity Localization and Tracking in GPS-Less Environments, MELT'09, pp. 235–254. Springer, Berlin (2009)
5. Gu, Y., Lo, A., Niemegeers, I.G.: A survey of indoor positioning systems for wireless personal networks. IEEE Commun. Surv. Tutor. **11**(1), 13–32
6. Baccour, N., Koubâa, A., Mottola, L., Zúñiga, M.A., Youssef, H., Boano, C.A., Alves, M.: Radio link quality estimation in wireless sensor networks: a survey. ACM Trans. Sen. Netw. **8**(4), 34:1–34:33 (2012)
7. Kouâa, A., Ben Jamâa, M., AlHaqbani, A.: An empirical analysis of the impact of RSS to distance mapping on localization in WSNs. In: Proceedings of The Third International Conference on Communications and Networking, ComNet 2012, pp. 1–7. IEEE (2012)
8. Sugano, M., Kawazoe, T., Ohta, Y., Murata, M.: Indoor localization system using RSSI measurement of wireless sensor network based on zigbee standard. In: Wireless and Optical Communications (2006)
9. Bahl, P., Padmanabhan, V.N.: RADAR: an in-building RF-based user location and tracking system. In: INFOCOM, pp. 775–784 (2000)
10. Lorincz, K., Welsh, M.: MoteTrack: a robust, decentralized approach to RF-based location tracking. Personal Ubiquitous Comput. **11**(6), 489–503 (2007)
11. Sree Divya, C., Soura, D., Zhi, D.: Distance estimation from received signal strength under log-normal shadowing: bias and variance. In: Proceedings of the 9th International Conference on Signal Processing, ICSP'08, pp. 256–259 (2008)
12. Patwari, N., Agrawal, P.: Calibration and measurement of signal strength for sensor localization. In: Guoqiang, M., Baris, F. (eds) Localization Algorithms and Strategies for Wireless Sensor Networks: Monitoring and Surveillance Techniques for Target Tracking, pp. 122–145 (2009)
13. Barsocchi, P., Lenzi, S., Chessa, S., Giunta, G.: Virtual calibration for RSSI-based indoor localization with IEEE 802.15.4. In: Proceedings of the 2009 IEEE International Conference on Communications, ICC'09, pp. 512–516. IEEE Press, Piscataway (2009)
14. Fu, Y.-J., Lee, T.-H., Chang, L.-h., Wang, T.-P.: A single mobile anchor localization scheme for wireless sensor networks. In: Proceedings of the 2011 IEEE International Conference on High Performance Computing and Communications, HPCC '11, pp. 946–950. IEEE Computer Society, Washington (2011)
15. Ji, Y., Biaz, S., Pandey, S., Agrawal, P., Ariadne: a dynamic indoor signal map construction and localization system In: Proceedings of the 4th International Conference on Mobile systems, Applications and Services, MobiSys '06, pp. 151–164. ACM, New York (2006)
16. Gwon, Y., Jain, R.: Error characteristics and calibration-free techniques for wireless LAN-based location estimation. In: Proceedings of the Second International Workshop on Mobility Management and Wireless Access Protocols, MobiWac'04, pp. 2–9. ACM, New York (2004)
17. Lim, H., Kung, L.C., Hou, J.C., Luo, H.: Zero-configuration, robust indoor localization: theory and experimentation. In: Proceedings of the 25th IEEE INFOCOM Conference, Barcelona (2006)
18. Dominguez-Duran, M., Claros, D., Urdiales, C., Coslado, F.: Dynamic calibration and zero configuration positioning system for WSN. In: The 14th IEEE MELECON Conference, Ajaccio (2008)
19. Varshavsky, A., Pankratov, D., Krumm, J., Lara, E.: Calibree: calibration-free localization using relative distance estimations. In: Proceedings of the 6th International Conference on Pervasive Computing, Sydney (2008)
20. Chintalapudi, K., Padmanabha Iyer, A., Padmanabhan, V.N.: Indoor localization without the pain. In: Proceedings of the 16th ACM MobiCom Conference, Chicago (2010)
21. Larranaga, J., Muguira, L., Lopez-Garde, J.-M., Vazquez, J.-I.: An environment adaptive zigbee-based indoor positioning algorithm. In: Proceedings of the IPIN Conference, Zurich (2010)

22. Bernardos, A.M., Casar, J.R., Tarro, P.: Real time calibration for RSS indoor positioning systems. In: Proceedings of the International Conference on Indoor Positioning and Indoor Navigation (IPIN), EEUU, IEEE (2010)
23. Wu, C., Yang, Z., Liu, Y., Xi, W.: WILL: wireless indoor localization without site survey. In: INFOCOM, pp. 64–72 (2012)
24. Fischler, M.A., Bolles, R.C.: Random sample consensus: a paradigm for model fitting with applications to image analysis and automated cartography. Commun. ACM **24**(6), 381–395 (1981)
25. Langendoen, K., Reijers, N.: Distributed localization in wireless sensor networks: a quantitative comparison. Comput. Netw. **43**, 499–518
26. Blumenthal, J., Grossmann, R., Golatowski, F., Timmermann, D.: Weighted centroid localization in zigbee-based sensor networks. In: IEEE WISP, Alcala de Henares (2007)
27. iLoc Overview: http://iloc.rtrackp.com (2012)
28. Savitzky, A., Golay, M.: Smoothing and differentiation of data by simplified least squares procedures. Anal. Chem. **36**, 1627–1639 (1964)
29. Tatham, B., Kunz, T.: Anchor node placement for localization in wireless sensor networks. In: WiMob, pp. 180–187 (2011)
30. Chen, Y., Yang, J., Trappe, W., Martin, R.P.: Impact of anchor placement and anchor selection on localization accuracy. In: Buehrer, M., (Reza) Zekavat, S.A. (eds.) Position Location—Theory, Practice and Advances: A Handbook for Engineers and Academics, Chap. 13, Wiley-IEEE (2011)
31. Rtrack project: http://www.rtrackp.com (2009)

Zeitfracht Medien GmbH
Ferdinand-Jühlke-Straße 7
99095 Erfurt, Deutschland
produktsicherheit@kolibri360.de